AF882198

Habtamu Haile Tolera

Suitability of Local Materials to Purify Akaki Sub-Basin Water

Karlsruher Mineralogische und Geochemische Hefte

Schriftenreihe des Instituts für Mineralogie und Geochemie,

Universität Karlsruhe (TH)

Band 33

Suitability of Local Materials to Purify Akaki Sub-Basin Water

von
Habtamu Haile Tolera

universitätsverlag karlsruhe

Dissertation, Universität Karlsruhe (TH),
Fakultät für Bauingenieur-, Geo- und Umweltwissenschaften, 2007
Referenten: Prof. Dr. Doris Stüben, Dr. rer. nat. Thomas Neumann

Anschrift des Autors:
Habtamu Haile Tolera
Klosterweg 28, A43
D – 76131 Karlsruhe

Anschrift der Schriftleitung:
Karlsruher Mineralogische und Geochemische Hefte
Institut für Mineralogie und Geochemie
Universität Karlsruhe (TH)
D – 76128 Karlsruhe

Impressum

Universitätsverlag Karlsruhe
c/o Universitätsbibliothek
Straße am Forum 2
D-76131 Karlsruhe
www.uvka.de

Universitätsverlag Karlsruhe 2007
Print on Demand

ISSN: 1618-2677
ISBN: 978-3-86644-151-4

Suitability of Local Materials to Purify Akaki Sub-Basin Water

Zur Erlangung des akademischen Grades eines

Doktors der Naturwissenschaften

an der Fakultät für Bauingenieur-, Geo- und Umweltwissenschaften
der
Universität Karlsruhe (TH)
genehmigte

DISSERTATION

von

Habtamu Haile Tolera

aus Addis Ababa (Äthiopien)

2007

Tag der mündlichen Prüfung: 14.02.2007

Referentin: Prof. Dr. Doris Stüben

Korreferent: Dr. rer. nat. Thomas Neumann

Zusammenfassung

Die Trinkwasserqualität in den südlichen Stadtrandgebieten der äthiopischen Hauptstadt Addis Ababa ist bis heute ein großes ungelöstes Problem. Ausgehend von der völlig unkontrollierten Ablagerung und Einleitung sowohl fester als auch flüssiger Abfälle in verschiedene Oberflächen Gehwässer stellt die Verschmutzung des Trink und Brauchwassers eine große Gefährdung der Gesundheit und des allgemeinen Wohlbefindens der dort lebenden Bevölkerung dar. Die wichtigsten Verursacher der Gewässerverunreinigungen im Akaki-Becken sind Gerbereien, Brauereien, Destillationsbetriebe, Marktplätze, Lebensmittelhersteller, Chemieindustrie, metallverarbeitende Betriebe, aber auch die landwirtschaftlich genutzten Flächen sowie wilde Hausmüllablagerungen. Die in dem Becken gelegenen ländlichen Siedlungen sind von den starken Verunreinigungen des Oberflächen und Grundwassers besonders betroffen. Durch Schwermetalle und andere anorganische Verbindungen wie beispielsweise Nitrat belastetes Trinkwasser ist für die hohe Anzahl von Krankheitsfällen in diesem Gebiet verantwortlich. Hier treten vor allem im Wasser gelöste, aber auch partikelgebundene Schwermetalle und andere Elemente wie Cu, Zn, Ni, As, Pb, Mn und Fe besonders häufig und unzulässigen Konzentration auf.

Diese Arbeit stellt einerseits die Ergebnisse einer hydrochemischen Charakterisierung von Oberflächen und Grundwasser aus Teilgebieten des Akaki-Beckens vor und bewertet andererseits ein innovatives Verfahren zur Entfernung von Schwermetallen aus entsprechend verunreinigtem Wasser unter Verwendung von einheimischen vulkanischen Gesteinen. Im April 2003 wurden Proben von Oberflächen und Grundwässern entnommen, um die Wasserbeschaffenheit in diesem Gebiet zu untersuchen. Die Probenahme vor Ort und auch die später im Labor durchgeführte Wasseranalytik erfolgte derart, dass der Transport von Schadstoffen in gelöster Form, aber auch in Verbindung mit Schwebstoffen und Kolloiden betrachtet werden konnte. Ausserdem wurden insgesamt zwölf Materialproben von einheimischen Gesteinen vulkanischen Ursprungs gesammelt. Davon wurden drei Proben für eine Detailstudie hinsichtlich ihrer Eignung zur Entfernung von Schwermetallen aus Wässern ausgewählt: DQ (Dika) Scoria, KVA (Kumo) Vulkanasche und ein lateritischer Ton der EBF (Ethiopian Brick Factory). Für die entsprechenden Versuche wurden zusätzlich und zu

Vergleichszwecken auch noch zwei Proben aus Indonesien eingesetzt: das Sediment MnR (Manganese Rich) sowie pyroclastisches Material PYC vom Vulkan Merapi.

Die hydrochemische Untersuchung der Wasserproben erbrachte, dass es sich hauptsächlich um erdalkalireiche Wässer handelt, wobei die Konzentrationen der Alkalien, und vorherrschend Hydrogenkarbonat, ebenfalls erhöht sind. Die gemessenen pH-Werte liegen zwischen 6,7 und 8,0. Die Konzentrationen von gelöstem Mn, Fe, Zn und Pb liegen bei den Grundwasserproben im Bereich der kritischen Grenzwerte der US EPA MCL, bei den Oberflächenwässern für Mn, Fe und Ni jedoch darüber. In allen untersuchten Wasserproben übertrafen die Konzentrationen für Uran den entsprechenden Grenzwert der US EPA deutlich. Ausgehend von der LAWA-Klassifikation als Bewertungsgrundlage wiesen alle Wasserproben ebenfalls erhöhte Konzentrationen von partikelgebundenen Schwermetallen auf. Vor allem die gemessenen Konzentrationen von Cu, Pb und Zn machen deutlich, dass die Grundwässer mäßig bis stark, die Oberflächenwässer dagegen „nur“ mäßig mit diesen Schwermetallen belastet sind. Untersuchungen von SPM mit dem Elektronenmikroskop (SEM) ergaben, dass sich die Schwebstoffe hauptsächlich aus Tonmineralen (vor allem Montmorillonit), organischen Partikeln und verschiedenen Eisenoxidverbindungen zusammensetzen. Die chemische und mineralogische Charakterisierung der verschiedenen Gesteine vulkanischen Ursprungs sowie des Laterits und MnR-Sediments ergaben, dass diese sich hinsichtlich einer Entfernung von Schwermetallen aus entsprechend verunreinigten Wässern tatsächlich als potentielle Filtermedien eignen, wobei allerdings ihre Rückhaltekapazitäten noch weiter bestimmt werden mussten.

Die entsprechenden Untersuchungen ergaben, dass die Reduzierung sowohl der gelösten als auch partikelgebundenen Gehalte von Cu, Zn, Ni, As und Pb im wesentlichen von zwei wichtigen Parametern abhängen. Im Labor wurden über einen Zeitraum von 12 Wochen hinweg Säulenversuche mit künstlich hergestellten verunreinigten Wässern durchgeführt. Diese Säulenversuche erfolgten im von oben nach unten gerichteten Durchflußmodus unter wassergesättigten Bedingungen und konstant bei pH = 7,0 und T = 20°C. Das Rückhaltevermögen der verschiedenen Filtermedien bezüglich der oben genannten Schwermetalle wurde miteinander verglichen. Die durchschnittliche Retention von Cu, Zn und Pb betrug bei allen Filtermaterialien über

90%. Bei Ni war sie jedoch in Abhängigkeit vom eingesetzten Filtermaterial sehr unterschiedlich (69 - 98%). Die Entfernung von As war mit 97% nur bei denjenigen Säulen sehr effektiv, die mit dem MnR-Sediment befüllt waren. Bei den Materialien DQ Scoria und dem lateritischen Ton EBF war innerhalb der ersten sechs Wochen eine Retention von immerhin noch 50% zu verzeichnen, während alle anderen Filtermedien ein nur sehr geringes Rückhaltevermögen besaßen. Die anschließend erfolgte Desorption von Mn und Fe war ebenfalls nur bei denjenigen Säulen vergleichsweise höher, die mit DQ Scoria und EBF befüllt waren. Das BCR-Extraktionsverfahren ermöglichte die Beobachtung, dass die Entfernung von Zn, Cu und Ni hauptsächlich auf einer schwachen Sorption an den Oberflächen der eingesetzten Filtermaterialien und/oder dem Montmorillonit beruht.

Insgesamt konnte mit dieser Arbeit hinreichend aufgezeigt werden, dass sich die untersuchten Gesteinsmaterialien vulkanischen Ursprungs sowie ihre Verwitterungsprodukte grundsätzlich als Filtermedien eignen. Es erscheint naheliegend, die Filtermedien auch in einem größeren (technischen) Versuchsmaßstab einzusetzen und weiter zu untersuchen.

Executive Summary

The drinking water status of the southern margins of Addis Ababa is a crucial problem at present day. Due to uncontrolled solid and liquid wastes discharged in to different open water systems, pollution of fresh water sources pose a great danger to the health and well being of the local people. The main polluting sources in Akaki Sub-basin are tanneries, breweries, distillers, open markets, food and chemical processing industries, steel and metal factories, agricultural fields and domestic wastes. In this basin, rural communities use polluted surface and groundwater for drinking and household purposes. Drinking water contaminated by heavy metals and some other inorganic pollutants such as nitrates account for the high rate of health hazards in this area. Dissolved and suspended particulate matter (SPM) bound Cu, Zn, Ni, As, Pb, Mn, and Fe are common pollutants in surface and groundwater.

This thesis (1) presented results of the hydrochemical characterization of surface and groundwater in parts of the Akaki Sub-basin and (2) evaluated an innovative approach for removing heavy metals from polluted water using local volcanic materials. In April 2003, surface and ground water samples were gathered to characterize the water quality in the area. The sampling and analytical efforts were focused on two phases (1) contaminants transported in the dissolved phase, and (2) those transported in association with the suspended particles and colloids. Twelve local materials of volcanic origin were also collected by grab method. Three materials DQ (Dika) Scoria, KVA (Kumo) Volcanic Ash, and EBF (Ethiopian Brick Factory) clay/laterites were selected for detailed study of heavy metals removal. MnR (Manganese Rich) sediment and PYC (Merapi) pyroclastic were also taken from Indonesia for comparison.

Results of hydrochemical analysis of the water samples showed that the water type is primarily of alkaline earth rich fresh water with considerable amounts of alkalis, prevailing hydrogen carbonate, with pH values ranging from 6.70 to 8.00. Dissolved concentrations of Mn, Fe, Zn, and Pb in groundwater were at the critical limit of the US EPA MCL while surface water showed Mn, Fe, and Ni exceeding the US EPA

MCL. In all of the water samples the concentration of U was well above the critical limit given by the US EPA. All water samples showed elevated concentrations of suspended particulate matter (SPM) bound heavy metals in relation to the LAWA classes. Comparison of SPM bound concentrations of Cu, Pb and Zn with the LAWA classes indicated that groundwater samples are moderately to strongly contaminated with the respective heavy metals. On the other hand, surface water samples are moderately contaminated with Cu, Pb, and Zn. As determined by Scanning Electron Microscopy (SEM), the main compositions of SPM are clay minerals (montmorillonite), organic mater and some oxides of Fe. Chemical and mineralogical characterization of the filter media showed that the volcanic materials, laterites and MnR sediment are potential filter media to further investigate their capacities in removing heavy metals from polluted water.

In order to purify the polluted water suitability of the filter media were examined. Two important parameters in this investigation include the reduction of SPM and the removal of dissolved Cu, Zn, Ni, As and Pb. In laboratory experiment simulated polluted water was studied for 12 weeks in column experiment. The column experiments were performed in a downward flow modus under saturated condition at pH =7, and T= 20°C. The removal capacities of the filter media for dissolved and SPM bound Cu, Zn, Ni, As, and Pb were compared. The average retention of Cu, Zn, and Pb by all filter media was above 90%. However, Ni retention was variable (69-98 %) depending on the filter substrate investigated. Removal of As was effected only in columns filled with MnR sediment (97%) while DQ scoria and EBF laterite retained 50% of As in the first 6 weeks. All the other media showed poor arsenic removal capacities. The desorption of Mn and Fe from columns filled with DQ and EBF was also higher compared to the other filter media. From BCR extraction procedure it was possible to observe that Zn, Cu and Ni removal mechanisms mainly involved weak sorption including surface precipitation on the media and/or on montmorillonite.

Overall, the project adequately demonstrated the potential applicability of volcanic materials and their weathering products as filter media, and it appears that the filter media are appropriate for testing at a larger scale.

አጭር መግለጫ

በአዲስ አበባና አከባቢው ያለው የመጠጥ ውሀ ሁኔታ በአሁኑ ወቅት በአሳሳቢ ደረጃ ላይ ይገኛል፡፡ ያለአግባብ ወደ አቅራቢያ ወደሚገኙ የውሀ ሀብቶች በሚወገዱ የደረቅና ፍሳሽ ቆሻሻ ምክንያት የሚበከለው ውሀ በአከባቢው ነዋሪዎች ላይ ትልቅ የጤና ጠንቅ እያስከተለ ይገኛል፡፡ በዋነኝነት ከሚጠቀሱ የብክለት ምክንያቶች ውስጥም የቆዳ ፋብሪከዎች፤የቢራ ፋብሪከዎች፤ክፍት የገበያ ቦታዎች፤የጨርቃጨርቅ ማምረቻዎች፤ የምግብና ኬሚከል አምራቾች፤የብረታ ብረት ፋብሪከዎች፤ከእርሻና ከቤት ውስጥ የሚወገዱ ፍሳሾች ይገኙበታል፡፡ በአቃቂና አከባቢው የወራጅና ከርሰ ምድር ውሃ ዋነኛ የመጠጥና የቤት ውስጥ አገልግሎት ምንጮች ናቸው፡፡ ይሁን እንጂ በሄቪ ሜታልና በናይትሬት በተበከሉ የወራጅና የከርሰ ምድር ወሀ ምክንያት ብዙ የጤና ችግር እየደረሰ ይገኛል፡፡ በውህደት ውሀ ውስጥ በሚንሳፈፉ የውሀ ውስጥ አከላት- (ኤ.ስ.ፒ.ኤም) የሚገኙ ኬሚከሎች ለምሳሌ Cu, Zn, Ni, As, Pb, Mn, እና Fe ከመጠጥ ውሀ ጋር አብረው ወደ ሰውነት ይገበሉ፡፡ ተመሳሳይ የወሀ ምንጭም ለቤት ውስጥ አገልግሎት ይውላል፡፡

ይህ ፅሁፍ 1) በአቃቂና አከባቢው ባሉ የመጠጥ ውሀ ምንጮች ላይ ያለውን ኬሚከላዊ ይዘትና 2) ከእነዚህም ውስጥ የተበከሉትን ውሀዎች በአከባቢው በሚገኙ የእሳተ ጎመራ ውጤት አፈርና ደንጊያ ስለ ማጣራት ሙከራና የተገኘውን ውጤት ያስገነዝባል፡፡ በሚያዚያ ወር 1996 ዓ.ም ከአቃቂና አከባቢው የወራጅ ወሀ፤የከርሰ ምድርና የፋብሪከ ፍሳሽ ውሀ ናሙናዎች ጀርመን አገር በቤተ ሙከራ ተጠንተዋል፡፡ የውሀ ምርምሩ በሁለት ጉዳዮች ላይ ያተኩራል፡፡ እነዚህም (1) በውህድ ውሀ ውስጥ የሚገኙ ኬሚከሎችንና (2) በኤ.ስ.ፒ.ኤም ላይ ተደብለው ያሉ ኬሚከሎችን ያከትታል፡፡

ይህንን የተበከለ ወሀ ለማጣራት ይረዳ ዘንድ 12 የእሳተ ጎመራ ውጤት አፈርና ደንጊያ ናሙና ከአቃቂና አከባቢው ተሰብስበው በቤተ ሙከራ ጥናት ተደርጎባቸዋል፡፡ ከእነዚህም ውስጥ የዲከ ስኮሪያ፤ኩሞ የእሳተ ጎመራ አመድ፤የኢትዮጵያ ሸክላ ፋብሪከ ሸክላ አፈር፤ይገኙባቸዋል፡፡ ተመሳሳይ የእሳተ ጎመራ አመድና በማንጋኒዝ የበለጸገ የሴድመንት ናሙና ከኢንዶኔዥያ ተወስዶ ለምርመራ አገልግሎት ውሏል ፡፡

የቤተ ሙከራው ጥናት ውጤት እንደሚያሳየው በአቃቂና አከባቢው ያለው የመጠጥ ውሀ ያልከላይን እርዝና መጠነኛ አልከሊስ ያዘለ፤ እንዲሁም በቂ ሀይድሮጅን ከረቦኔት መጠን ያለው ሲሆን የ ፒ. ኤች መጠኑም በ 6.7 እና 8.0 መከከል ይገኛል፡፡ ሰማንያ

ሰባት በመቶ የሚሆነው የከርሰ ምድር ውሃ ውስጥ ያለው የናይትሬት መጠን ያሜሪክን የአከባቢ ጥበቃ ኤጀንሲ ከወሰነው መጠን (10 mg/l) በላይ የላቀ ነው። በከርሰ ምድር ውሀ በውህደት ውስጥ ያለው የ Mn, Fe, Zn, ና Pb መጠንም እንደዚሁ ያሜሪክን የአከባቢ ጥበቃ ኤጀንሲ ለመጠጥ ውሀ ከወሰነው መጠን የላቀ ነው። በሌላ በኩል ደግሞ በወራጅ ወሀ ውስጥ ያለው Mn, Fe, እና Ni መጠን ለመጠጥ አገልግሎት በሚውሉ ውሀ ውስጥ በከፍተኛነት ተመዝግቧል። በውህደት ውስጥ ያለው የ U መጠን በአጠቃላይ በሁሉም የውሀ ናሙናዎች ያሜሪክን የአከባቢ ጥበቃ ኤጀንሲ ከወሰነው መጠን (0.2 µg/l) በላይ የላቀ ነው።

በኤ.ስ.ፒ.ኤም ላይ ያለው ሄቪ ሜታል ይዞታም ከ LAWA የአውሮፓ የአከባቢ ጥበቃ ኤጀንሲ ከወሰነው የኤ.ስ.ፒ.ኤም ብክለት አመልካች ጣሪያ በላይ ነው። ከዚህም አንጻር በከርሰ ምድር ውሀ ኤ.ስ.ፒ.ኤም ላይ ተደብለው ያሉ Cu, Zn እና Pb መጠን እንደሚያመለክተው የከርሰ ምድር ውሀ በከፍተኛ ሁኔታ መበከሉን ነው። በሌላ በኩል ደግሞ በወራጅ ወሀ ውስጥ ያለው የኤ.ስ.ፒ.ኤም Cu, Pb, ና Zn መጠንም መከከለኛ የሆነ ብክለትን ያመለክታል። የኤሌክትሮን ማጉያ መነጽር መሳሪያ እንደሚያሳየው ከሆነ የኤ.ስ.ፒ.ኤም ውህደታዊ ይዘት የሸክላ አፈር፣ ህይወት ያላቸው ነገሮች ብስባሽ ቅሪት፣ እና አንዳንድ የአይረን ኦክሳየድ ውህደቶችን ያሳያል። በእሳተ ጎመራ ውጤት አፈርና ደንጊያ ላይ የተደረገው ኬሚከላዊና የማእድናት ወሀደታዊ ይዞታ ጥናት እንደሚያሳየው ከሆነ ደግሞ እነዚህ አፈርና ደንጊያዎች በሄቪ ሜታል የተበከለ ወሀን ለማጣራት ከፍተኛ አቅም እንዳላቸው ያመላክታሉ።

ይህንንም ለማረጋገጥ ይረዳ ዘንድ የአፈርና ደንጊያ ናሙናዎቹ ለ 12 ሳምንታት በቤተ ሙከራ የኮለም ጥናት ተደርጎባቸዋል። የኮለም ጥናቱም በመሰረታዊ የቤተ ሙከራ ሁኔታዎች (ፒ. ኤች 7.0 እና የሙቀት መጠን 20 ዲግሪ ሴንቲ ገሬድ) ላይ የተመረኮዘ ነው። በዚህም ሙከራ ዋነኛ ጥናት የተደረገባቸው ሁለት አብይ ፓራሜትሮች ኤ.ስ.ፒ.ኤምን ከውሀ ውስጥ ማስወገድና በውህድ ውሀ ውስጥ የሚገኙ Cu, Zn, Ni, እና Pb መጠንን ከውሀ ውስጥ ስለመቀነስ ያከትታል። የአፈርና ደንጊያ ናሙናዎቹ የተበከለ ውሀን የማጣራት አቅም እንደሚያሳየው በተጣራው ውሀ ውስጥ ያለው የውህድ Cu, Zn, እና Pb መጠን 90% መቀነሱን ያመለክታል። ይሁን እንጂ የ Ni መወገድ ሁኔታ የተለያየ መጠንን (69-98%) ያሳያል። የ As በአፈርና ደንጊያ ናሙናዎቹ መወገድ ግን በከፍተኛ ሁኔታ ወጤት ያሳየው በማንጋኒዝ የበለጸገ የሴድመንት ናሙና(97%) ወስጥ ሲሆን የዲከ ስኮሪያና የኢትዮጵያ ሸክላ ፋብሪከ ሸክላ አፈርም እስከ 50% የሚደርሰውን

As ለማስወገድ ችለዋል:: በሌላ በኩል ደግሞ የ Mn እና Fe ከኮለሙ አፈር ናሙናዎቹ አየታጠበ መወጣት ሌላው ሁኔታ ነው:: ይህም ሁኔታ በዲከ ስኮሪያና የኢትዮጵያ ሸክላ ፋብሪከ አፈሮች ወስጥ በከፍተኛ ሁኔታ ተመዘቧል::

በአጠቃላይ ሲታይ የኮለም ጥናቱ እንደሚያስገነዝበው በአቃቂ አከባቢ የሚገኙ የተፈጥሮ የእሳተ ጎመራ ውጤት አፈርና ደንጊያ በሄቪ ሜታልና በአንዳንድ ኢንኦርጋኒክ ኬሚከሎች የተበከለ ወሀን ለማጣራት በቀላሉ ጥቅም ላይ መዋል ይችላሉ:: በአከባቢው የሚገኙ ነዋሪዎችም ይህንን የጥናት ውጤት በቀላሉ ያለተጨማሪ ወጪ በተግባር ላይ ማዋል ይችላሉ::

Acknowledgements

This Project was funded by LGFG (Landesgraduiertenförderungsgesetz) and KAAD (Katholisher Akademischer Ausländer-Dienst). I am grateful to Professor Doris Stueben and Dr. Thomas Neumann for their invaluable support in organizing this PhD project, the field trip to Ethiopia and their active participation in the fieldwork. I would also like to thank the scientific and laboratory stuffs of IMG :Utz Krammer and Claudia Mößner for their support in the lab and analyzing the surface and groundwater samples taken from the Akaki Sub - basin (April 2003) by ICP MS and TRFA: Miss Gesine Preuß, Mrs Beate Oetzel, Mr Kristian Nikoloski, and Mr Predrag Zrinjscak for their support in sample preparation and analytical measurements during this program: Mrs Bender, Mrs. Tannhäuser, and Mr Nytz and other stuffs of the Institute of Mineralogy and Geochemistry (University of Karlsruhe) for their friendly support during my stay at the institute.

Furthermore, I would like to thank all colleagues whom I came across during my stay in this vibrant technology university. Special thanks go to my girl friend Karin Haugg who has given me the strength and inspiration during this work.

I would also like to extend my gratitude to members of the Earth and Geosciences Department, Addis Ababa University particularly Dr. Bekele, A., Dr. Ayenew, T., and Dr. Alemayehu, T. for their support and consultations during my field trip to Ethiopia. I would like to thank Dr. B. Hundammer, Department of Chemistry, Addis Ababa University for providing me valuable information and lab facility during the field trip. I would also like to thank Abebe, M. and his colleagues of the Ethiopian Science and Technology Commission for the data and information they provided me regarding the study area.

My special thanks go to my beloved parents who have been encouraging me all through my academic and professional career and the love they have given me through my entire life. I also wish to express my gratitude for my brothers, sisters and friends.

Table of Contents

List of Figures

List of Tables

List of Appendices

Abbreviations

AAWSA	Addis Ababa Water Sewerage Authority
ANC	Acid Neutralizing Capacity
a.s.l.	above seal level
BCR	Community Bureau of Reference
BOD	Biological Oxygen Demand
CAC	Commercial Activated Carbon
CCA	Chromated Copper Arsenate
CEC	Cation Exchange Capacity
COD	Chemical Oxygen Demand
CSA	Carbon Sulfur Analyzer
CWA	Carbon Water Analyzer
DO	Dissolved Oxygen
DOC	Dissolved Organic Carbon
DQ	Dika
EBF	Ethiopian Brick Factory
EEPA	Ethiopian Environmental Protection Energy
EEPCO	Ethiopian Electric Power Corporation
EDTA	Ethylenediamine tetraacetic acid
EDXRF	Energy-Dispersive X-ray Fluorescence spectrometry
EF	Enrichment Factor
ELICO	Ethiopian-Leather Industry PLC
FAAS	Flame Atomic Absorption Spectrometry
HPS	High Purity Standard
HR ICP MS	High Resolution Inductively Coupled Plasma Mass Spectrometry
HSAB	Hard-Soft Acid Base
IC	Ion Chromatography
IMG	Institute of Mineralogy and Geochemistry
IUPAC	International Union of Pure and Applied Chemistry
K_d	Distribution Coefficient
K_{sp}	Solubility Product
KVA	Kumo Road Volcanic Ash

LAWA	Bund/Länder-Arbeitsgemeinschaft Wasser (The German Working Group on Water Issues)
MCL	Maximum Contaminant Limit
MERS	Main Ethiopian Rift System
MnR	Manganese Rich
NOM	Natural Organic Matter
NRC	National Research Council
OECD	Organization for Economic Co-operation and Development
PE	Polyethylene
PH_{PZC}	pH at Point of Zero Charge
ppb	Parts Per Billion
ppm	Parts Per Million
PYC	Pyroclastic
QTZ	Quartz
SEM	Scanning Electron Microscopy
SMCL	Secondary Maximum Contaminant Limit
SOM	Soil Organic Matter
SPM	Suspended Particulate Matter
SPW	Simulated Polluted Water
TDS	Total Dissolved Solids
TMDW	Trace Metal in Drinking Water
TXRF	Total Reflection X-ray Fluorescence
UNESCO	United Nations Educational, Scientific and Cultural Organization
USDA	United States Department of Agriculture
US EPA MCLUS	Environmental Protection Agency Maximum Contaminant Level
WDXRF	Wave-Dispersive X-ray Fluorescence Spectrometry
WHO	World Health Organization
XRD	X-Ray Diffraction
XRF	X-Ray Fluorescence

1 Introduction

Since decades water quality degradation has been one of the critical environmental problems in Addis Ababa and the surrounding. The fast population growth, the expansion of urbanization and industrialization, which are targeted in certain peripheral parts of the city, poor sanitation conditions, uncontrolled waste disposal and poor awareness of environmental issues pose serious health problems on the well being of the society. Contamination of drinking water sources in the southern margins of Addis Ababa is one of the critical environmental issues that need practical solutions and measures.

The southern margin of Addis Ababa is one example of the most polluted fresh water sources in Ethiopia. In this catchement most of the water resources for domestic, irrigation and industrial purposes are surface water and groundwater. Three important sources of surface water are identified in the southern margins of Addis Ababa, which are used by the local people: The Big Akaki River, the Small Akaki River and the Aba Samuel Lake.

Due to uncontrolled solid and liquid wastes discharged in to these open water systems, pollution of the water sources pose great danger to the health and well being of the local people. The environmental impacts of urbanization on the release of heavy metals in river water and sediments are well presented in the works of Liston and Maher (1986), Sarmani et al. (1992), Murray (1996), Singh et al. (1997), Fuchs et al. (1997), Estèbe et al. (1998). The effects on lake sediments were published by Striegl (1987), Nriagu et al. (1996), and Chillrud et al. (1999). The removal of heavy metals in urban runoff by sorption on mulch was found to be efficient (Jang et al. 2005). Due to the poor water management and waste disposal system toxic heavy metals, organic pollutants and bacteria heavily pollute waters.

Currently, the water quality problem in Addis Ababa has directly impacted the population living in the southern margins of the city where there is no municipal water supply. In the rural parts of Akaki where there is a serious shortage of municipal water supply, communities rely directly on untreated water from springs, ponds and other open water systems. Most of the springs are not developed and therefore are vulnerable to different types of pollutants. The local people in these communities use water from the Big and Small Akaki Rivers for drinking and other household purposes. However, these rivers are the major targets for discharging industrial effluents as well as private septic tanks. Domestic animals drink freely from the polluted rivers of Akaki and the Aba Samuel Lake. Irrigation of agricultural fields with water drawn from industrial wastes to produce vegetables has been practiced for decades. Though, shallow groundwater wells drilled close to the Akaki Rivers provide the maximum expected yield, their quality is highly degraded by coliforms, inorganic and organic pollutants (Alemayehu 2001).

Especially the major industries concentrated in the southern margin of Addis Ababa are accountable for the water pollution in the area. These include tanneries, breweries, wineries, distillers, soft drink, chemical and metal work industries, the Addis Ababa Abattoir Enterprise and the National Tobacco Factory. Municipal drain water including latrines goes directly to surface water (rivers, lakes, and open water systems). Water sludge from small-scale industries and factories are disposed without any treatment in the surface waters or dumped on open fields. Industrial waste water, irrigation runoffs, and domestic waste water diverted to public land and to the fresh water resources contain toxic substances and high content of total dissolved solids.

So far little has been done in alternative measures that will help improving the water quality that the rural communities of Akaki are relying on. It has been observed that surface waters are directly tapped from their sources and used with out any purification or pre-treatment (Alemayehu 2001). The local people are more concerned on the color and odor of the water sources. Particularly, the overall inorganic as well as organic constituents of these waters are not considered before using the water for drinking.

The contaminants of interest are dissolved as well as suspended particulate matter (SPM) bound Mn, Fe, Ni, Cu, Zn, As, Pb and Cd. In this regard, there is no previous research work that have been done in characterizing the distribution of some heavy metals and arsenic on suspended particulate matters of surface water and groundwater. In this particularly new approach the following objectives are included:

- Characterization of the contaminated waters in the Akaki area and find out the contamination levels in the vicinity of the main water resources.
- Mineralogical and chemical characterization of potential local filter materials.
- Laboratory column and batch experiments to study the metal removal capacities of local materials derived from volcanic deposits.

There have been several researches done in the removal of metals from drinking water and wastewater. Adsorption using commercial activated carbon (CAC) can remove heavy metals from wastewater, such as Cd (Ramos et al. 1997), Cr (Ouki et al. 1997), Ni (Shim et al. 2001), Cu (Monser and Adhoum 2002). In general, the conventional water treatment technologies applied in municipal drinking water supply are too expensive and their applications would be more complex for rural people. Therefore, there is a need to look into alternatives to investigate a low-cost treatment method that is effective and easily applicable.

The idea of applying low cost materials for the removal of toxic metals from polluted waters and/or wastewaters has been proved to be a better alternative over the conventional treatment methods (Bailey et al. 1998). Giridih coal and crushed coconut shell were found to adsorb cadmium from solution (Bhattacharya et al 1984). Adsorption of metals to filter materials has been studied using low-cost sorbents including carbonaceous materials, waste by-products, weathered soils, and agricultural/natural products such as activated carbon, iron-oxide-coated sand, zeolites, rice bran and hulls, soybean, peanut husk, saw dust, peat and leaves (Naidu et al. 1998, Brown et al. 2000a, Brown et al. 2000b, Ricordel et al. 2001, Färm 2002 and Ho et al. 2002). Geochemical process such as adsorption, co-precipitation, precipitation, ion exchange and redox reactions are mechanisms that are involved in the removal of heavy metals from contaminated water. The application of natural media in treating polluted water involves one or combination of these removal processes. Natural materials such as those tested in this work have shown positive outcome and have practical as well as economical importance from the point of this project. The potential positive out come of this investigation can also be adopted nationally in areas where geogenic and anthropogenic heavy metal pollution of drinking surface water and groundwater are critical problems.

2 The Akaki Sub-Basin

2.1 Geography and Climate

The Akaki Sub-Basin is located in the southern part of Addis Ababa between 38°41' - 38°49'E and 08°47' - 08°70'N. It is part of the Akaki river catchment area, which is represented by the city of Addis Ababa and the Akaki well fields (Fig. 2.1). The Akaki catchment covers a total area of 11,500 km^2 and is characterized by a sharp elevation drop of more than 1000 m towards the Great East African Rift System over a distance of 30 km. Compared to the low lying southern part of the Akaki catchment, the northern part of the catchment is characterized by ragged topography and steep slopes, which resulted in rapids and a high runoff coefficient.

The lower lying part of the Akaki catchment area extends southwards up to the Lake Aba Samuel, which lies at about 2000 m above sea level (a.s.l.) Contrasting topography such as the high mountain range of Furi (2839 m a.s.l.) represents the highest elevation in the area. The eastern part of the water shed is bounded by mountains ranging in elevations of 2400 - 2500 m a.s.l. (the highest Mt. Range in the area being Mt Yerer, 3100 m a.s.l.).

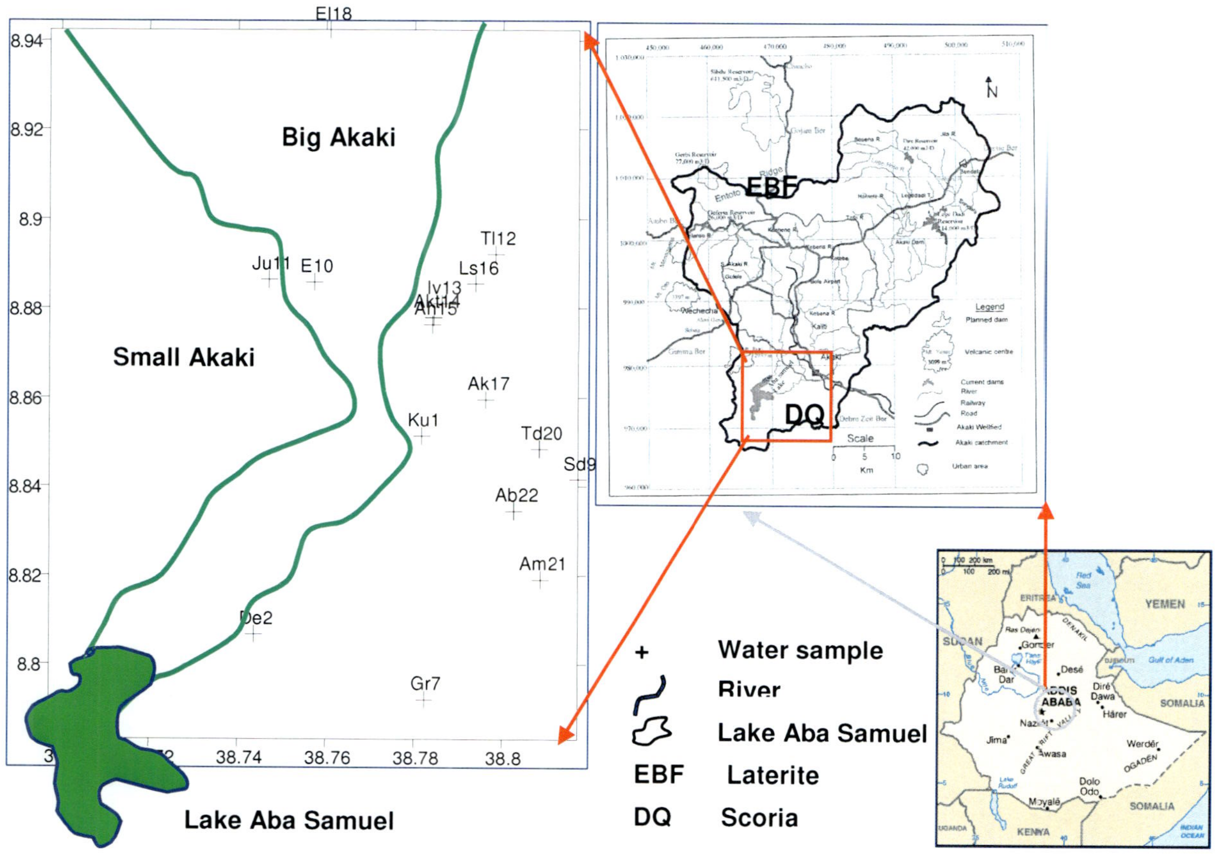

Fig 2.1. The Akaki catchment area includes Addis Ababa, the four artificial reservoirs and natural lakes.

Because of its favorable morphology the southern part of Addis Ababa is the fastest growing part of the city where several industries are densely established. According to the Ethiopian Central Statistics Authority (1999), 65% of the total industries in the country are situated in Addis Ababa.

According to the National Atlas of Ethiopia (1981), five traditional climatic zones have been drawn in the country. These are: "Kur" (Alpine), 3000 m and above; "Dega" (temperate), 2300 m to about 3000 m; "Weina Dega" (Sub tropical), 1500 to about 2300 m; "Kolla" (Tropical), 800m to about 1500 m and "Bereha" (Desert), less than 800m.

The Akaki Sub-Basin, which is part of Addis Ababa, is located in the region where the rainy months are contiguously distributed (Daniel 1977). In this basin, there are seven rainy months from March to September with intermittent rainy season occurring from March to May. The heavy rains are from June to September. High concentration of rainfall usually occurs in July and very high concentration in August. June, July, August and September represent the wet seasons of Addis Ababa and most parts of the highlands in the country.

The average annual rainfall that was measured at five different rain gauge stations within the Akaki catchment was 1250 mm (Ethiopian Meteorology Authority, 1981-1999). The highest temperature varies on average between 20 °C in the wet season and 25 °C during the dry season. The relative humidity of Addis Ababa seems to be minimal during the dry season (70%) while during the rainy season the values rise up to 90%. Wind speed is generally moderate in the area reaching up to 0.9 m/s. Average daily sunshine hours in the area reach as high as 9.5 hours during the months of November and December (Bega or dry season) while during the months of July and August this value falls to about 3 hours. Monthly evaporation records obtained from the Ethiopian Meteorology Authority showed that the average pan evaporation during the dry season (November) is about 180 mm and in wet season the values goes down to 75 mm. Evaporation in the low lying Akaki areas are relatively higher than the northern mountain ranges of Entoto areas.

2.2 Geology and Hydrogeology

2.2.1 Regional Geology of the Akaki Catchment

Wide ranges of volcanic rocks of different ages characterize the area that covers the catchment of Akaki River including Addis Ababa and the well field (Tsehayu and Hailemariam 1990, AAWSA and Seureca 1991, Morton et al 1979, Vernier and Chernet 1985). The rocks in Addis Ababa are highly affected by the rift tectonics that are very common features of the Ethiopian Rift Valley. The north and northeastern area (the Entoto Mountain, the northern and north eastern Addis Ababa) are covered with trachytes, rhyolites, basalts and several episodes of pyroclastic materials of older volcanism occur on the upper part and foothill sides of Entoto ridge. The central and southern part of the Addis Ababa area is overlain by younger basaltic rocks (7 - 6.4 My) (Addis Ababa Basalt). Outcrops of ignimbrites north of Bole area (Eastern Addis Ababa) and Lideta area (Central Addis Ababa) have been observed underlying the Addis Ababa basalt. Younger volcanic material of trachy-basalt, trachytes, ignimbrites and tuff belonging to the Wochecha, Furi and Yerer volcanoes are recognized overlying un-conformably on the Addis Ababa basalt in the western, southwestern and eastern part of the catchment. Akaki (including the well field), Dukem and Debre Zeit area are covered with olivine basalt, scoria, scoriaceous basalts and vesicular basalts. In the upper part, in some places, intercalation of tuffs, sand and gravels are also recognized. In northeast, east and in smaller extent in the western part of Debre Zeit, outcrops of trachytes, rhyolitic ignimbrites and tuffs are recognized covering the area (Fig. 2.2).

The overburden covering the bedrocks includes soils, lacustrine & alluvial deposits. In the central, southeastern, northeastern, western (Kolfie & Keraniyo) and in northern (Gullelie) part of the catchment thick in-situ soil type cover is found. The lacustrine covers Bole, Lideta, Mekanisa, Akaki-Aba Samuel area and Dukem-Debre Zeit area. Along Big & small Akaki Rivers, in the southern and southwestern parts of Addis Ababa, some alluvial deposits occur and minor deposits are also present along Kebena River in the area northwest of Bole (Tadesse 2004).

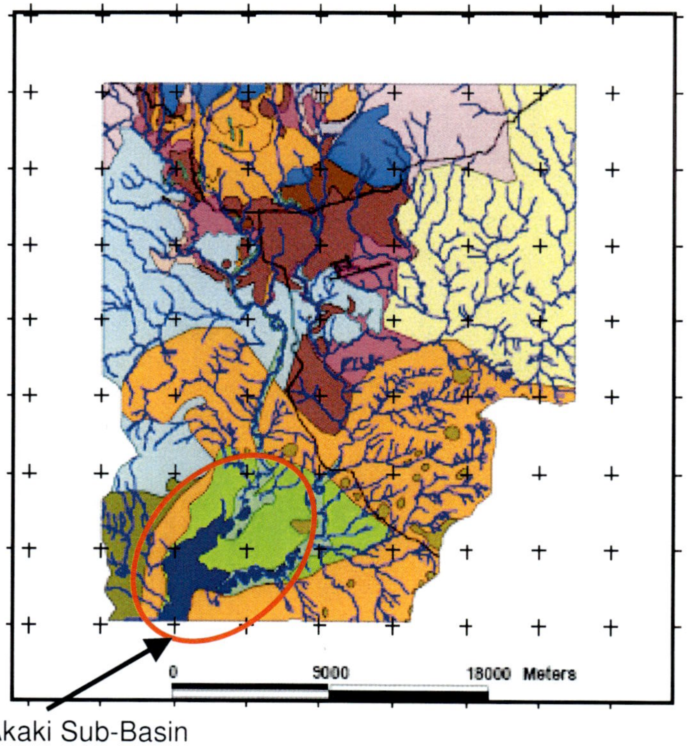

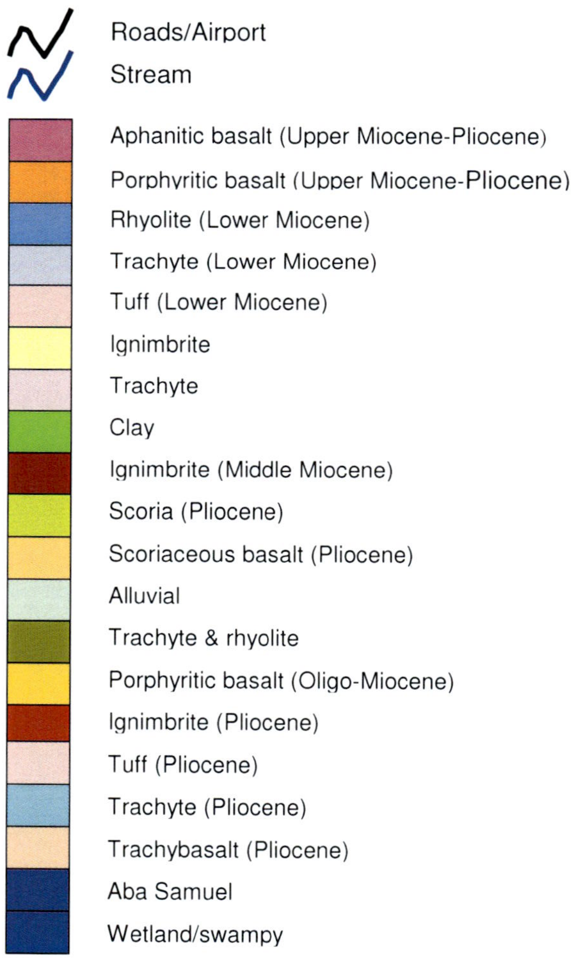

Fig. 2.2. Geological Map of the Akaki Catchment (Girmay et al 1989).

Some of the most common tectonic features characterizing the Akaki catchment include the Main Ethiopian Rift System (MERS). The Akaki catchment lies at the upper western margins of the Main Ethiopian Rift System. Therefore, it is highly affected by rift tectonics, which is manifested by numerous faults having similar trend as that of the rift (NE-SW), and faults and lineaments of E-W, N-W and NE-SE trend.

Volcanic rocks and minor alluvial sediments that occur along the banks of both Big and Little Akaki Rivers are the main aquifer systems where the groundwater occurs in Akaki catchment. These volcanic aquifers have primary porosities such as vesicles and joints while faults, fractures and fissures constitute the secondary porosities. Groundwater circulation and occurrence in these volcanic rocks are associated with these porosities while in the alluvial aquifers they are in the interstitial spaces between the sediments. In the catchment area the geometry of the aquifer systems is highly variable, discontinuous and not well defined. In most areas, the volcanic aquifers show semi-confined to unconfined natures and in few areas (Kerchelie & Kality) confined aquifers are common.

North of Akaki Bridge there is interaction between groundwater with surface water of Big and Little Akaki Rivers. Consequently, the base flow of the rivers is contributed from the groundwater. The recharge to the groundwater that takes place within the Akaki catchment to the north of Akaki Bridge is considered contributing to the base flow. The piezometric surface constructed from groundwater points inventory made during previous projects showed that the general groundwater flow direction is from north to south in the upper and central part and towards south & southeast in the lower parts of the catchment area (Tadesse 2004).

Most soils in this area are highly eroded and form thin soil cover. Most soil profiles are products of weathering of old basic and acidic rocks that outcrop in the central, western, and southwestern parts of Addis Ababa. The detrital materials that are derived from elevated area of Entoto, Wechecha, Furi and Yerer are transported and deposited in the pedimont and along the stream courses of Addis Ababa. Soils are black in color and their thickness varies from place to place primarily depending on the slope of the area. Clay is the dominant composition of the soils in the Akaki area. The thickness of the soils varies from 1 m to 20 m around the Akaki Well field.

The alluvial sediments cover large parts of the Akaki River area along riverbanks. Along the Akaki River and the surrounding the thickness of the exposed soil cover reached up to 5 m. Grain sizes are variable ranging from sand to gravel showing poor sorting. The alluvial sediments are characterized by high porosity and intergranular permeability. These sediments constitute one of the potential aquifers in the Akaki area.

2.3 Hydrology

The city of Addis Ababa, the Akaki River and four surface water reservoirs namely Gefersa, Lege Dadi, Dire Dam, and the Lake Aba Samuel, which are located in the NW, NE, N, and SW of Addis Ababa respectively are the most important hydrological features of the Akaki catchment. These are the major surface water bodies within and around the city of Addis Ababa. All the reservoirs have been constructed for Addis Ababa water supply except the Lake Aba Samuel. The reservoir is a waste-damping site that collects solid and liquid wastes that drain through various parts of the city of Addis Ababa.

Several small and relatively big rivers and streams that drain the different parts of Addis Ababa characterize the Akaki River catchment (Fig. 2.3). The Akaki River that constitutes the largest part of the catchment area drains the city of Addis Ababa from the northeastern and eastern direction. Commonly this river is known as the Big Akaki River. The other equally significant river that constitutes the Akaki catchment area is the Little Akaki River that drains the western part of the catchment area. The upper courses of these rivers originate from the northern mountain ridges of the catchment area and enter in to Lake Aba Samuel at the southern reach (Fig. 2.3). The Aba Samuel reservoir was built on the Akaki River, 30 km south of Addis Ababa. The dam has a storage capacity of 65 million cubic meters and an annual out put of 23 million kilowatt. However, due to siltation and pollution it is not functional at present. On the other hand, Gefersa, Legedadi and Dire Dams are located in the upstream part of the Akaki catchment are used as water supply sources of Addis Ababa. The Akaki River flows further down through a gorge of 100 m deep for about 8 km before joining the Awash River, one of the biggest rivers in Ethiopia that has been utilized for irrigation.

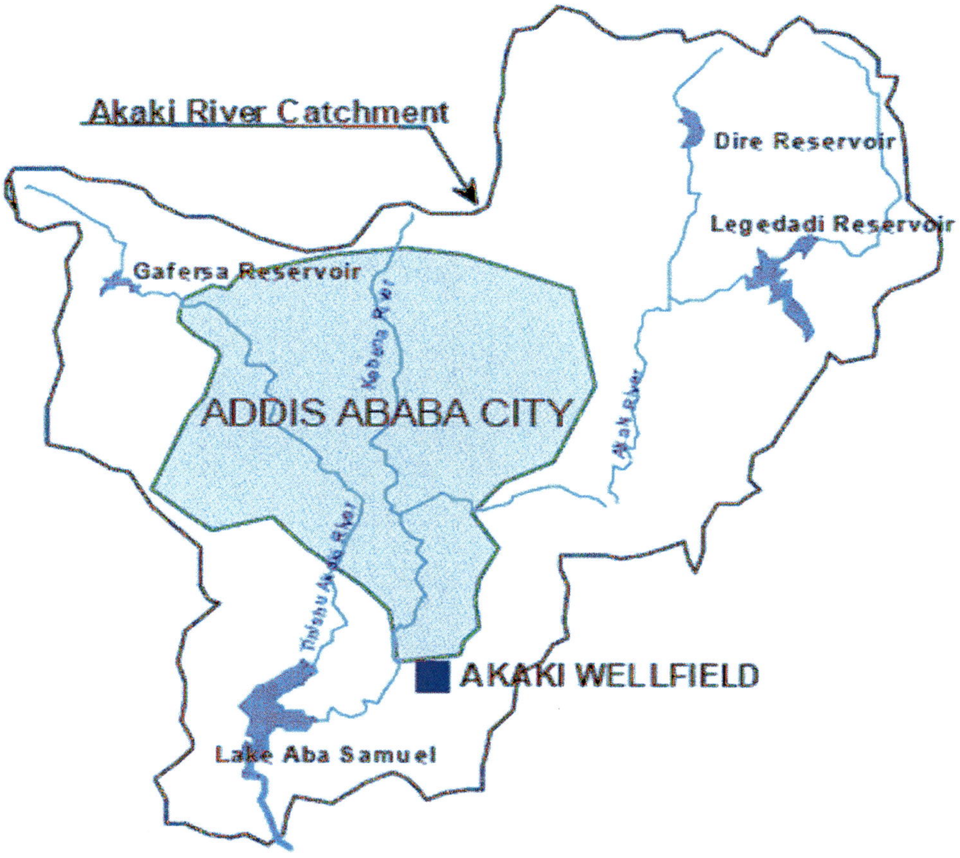

Fig. 2.3. The Akaki catchment area includes Addis Ababa, the four artificial reservoirs.

Several tributaries that flow through the eastern and northeastern part of Addis Ababa join the Big Akaki River. These tributaries include Ginifle, Kebena, Kechene, Kurtume and Yeka, which are all found within the eastern part of the city boundaries. The Big Akaki River loses some 600 m water head throughout its course before it confluents with the Awash River near Dodota, 1880 m a.s.l. (AAWSA et al. 2000).

The Akaki well field is one of the most important groundwater protection zones in the south outskirt of Addis Ababa. It lies in the lower part of the Akaki River catchment within the drainage basin of the streams Dengora and Keta, which join to form the Sakelo. The Sakelo River in turn joins the Big Akaki River. The Akaki well field lies in a topographic plain with an elevation of about 2100 m a.s.l. and has a flow direction in general towards southwesterly to the Aba Samuel Lake (2020 m and 2025 m a.s.l.).

North south running surface water divide separates the total catchment area of the Akaki River basin. In the catchment area, the streams drain towards south from the Entoto ridge; southeast direction from Mt. Wechecha and Mt. Furi and towards southwest direction from Mt. Yerer and other elevated areas of the eastern outskirts of the city. Some perennial streams in the city are part of the catchment such as Little Akaki, Bantyiktu, Kurtume, Kebena, Ginfile, and Big Akaki. Other streams are intermittent in nature.

At the northern sections, streams are dense forming characteristic radial drainage pattern while on the slope and most parts of the city dendritic drainage patterns are common. In addition to this, the steep slopes facilitate the processes of erosion that result in cutting sharp valley floors. On the other hand, in the center and southern parts of the city, the density of the streams is low and the main rivers or big tributaries show a wide meandering type of flow. Low lying topography in the southern margins creates fewer gradients whereby the streams are flowing in meandering course. Formations of alluvial deposits in the southern margins are due to the decrease in gradient of the topography, lower eroding activity of the rivers, minimum flow velocity and transporting capacity (AAWSA, AAU 2003).

In the study area rivers and/or tributaries are located near the watershed and find their sources from the steeply and rugged ridges of the northern mountain ranges. Most of the streams flow in rapid movement through the city of Addis Ababa towards the low-lying areas of the southern margins. Hence, the amount and volume of water flowing in the streams at the southern margins of Addis Ababa increases after heavy rainfall. In general, natural as well as man-induced features have affected the magnitude of peak flow in the southern margins of Addis Ababa. Flooding of the Akaki catchment area has been happening at certain periods of the year for several years. The Akaki areas are also just part of this frequent flooding resulted in several losses of properties. (AAWSA, AAU 2003).

2.4 Water Quality and Use

The main land use features in the city of Addis Ababa and Akaki areas include residential areas, open market quarters, industrial areas, small scale agricultural fields, forest covers, meadows and quarries. Without paying much attention to the environmental impacts solid and liquid wastes are discharged in to unprotected open areas where rivers and other open water systems are used for multi purpose household and agricultural practices.

Most industries, small-scale farmers and factories target their activities on the geographically low lying Akaki areas, along river channels, and the outskirts of Addis Ababa. One of the major polluting sources in the city believed to be the open waste dumping site (0.3 km^2) that is located in the western part of Addis Ababa. The landfill site has not been constructed in such a proper way that it may prevent infiltration of contaminants in to the nearby surface water and groundwater (Alemayheu 2001).

Regarding water quality, the Akaki River is one of the most contaminated rivers in the Akaki area mainly by industrial effluents. In the Akaki River just at the inlet to the Aba Samuel Lake biological oxygen demand (BOD) values up to 13 mg/l has been reported (Alemayehu 2001). On the other hand, samples taken from the reservoir outlet shows lower values of BOD (0.8 mg/l). However, the lake is used for drinking (humans and animals), irrigation and household activities (Table 4.1). The other important source of water in the area is Little Akaki, which carries the most toxic organic, inorganic and biological contaminants that are discharged from households, small industries, factories and open market places. Since the river drains through the most populated part of the city, its contamination load is high with dissolved oxygen concentration dropping down to $\approx$ zero in Aba Samuel Lake (Alemayehu 2001). In spite of the high contamination level, small communities downstream of the rivers utilize the river water as their potable water source. Most communities around the Aba Samuel Lake depend on water wells and sometimes on the most polluted rivers and the reservoir as their drinking water supply. Personal communications and field observations showed that, the community at village Dika uses water well in the village for drinking while the lake water is used for household purposes, drinking for domestic animals and fishing. At extreme dry conditions, the villagers rely on the most polluted waters of the Aba Samuel Lake. Some of the drinking water wells are located few hundred meters away from the lake. In addition to this, most of the ground water wells are located close to the Akaki River systems in order to get the maximum yield possible. Depending on the lithology and structural features in the subsurface, there is a potential groundwater-surface water interaction between the reservoir and groundwater, and the river systems and groundwater.

More than thousands of rural people also intensively use groundwater for drinking and household purposes. Wastewater discharged by the Akaki Textile Factory in to the nearby agricultural field is also part of the pollution problem. Farmers in the area directly use this water for irrigation to grow vegetables. It has been reported that leafy vegetables that have been irrigated with liquid waste released from this factory surpassed the expected tolerable limits of heavy metals (Itanna 1998)

There have been relatively very few published researches in this very critical environmental pollution issue of the present day Ethiopia. The application of industrial liquid waste in agricultural field happened to be the most common practices in areas where water resources are scarce.

Water analysis on the Big Akaki River water samples has shown higher amount of BOD at the inlet to the Aba Samuel Lake. However, this amount dropped at the outlet of the reservoir indicating that the lake acts as an oxidation chamber in that the water regenerates its dissolved oxygen content (Alemayheu 2001). The dissolved oxygen (DO), biological oxygen demand (BOD) and chemical oxygen demand (COD) for the Little Akaki Basin, Big Akaki Basin, and some tributaries of these basins are given in table 2.1.

Table 2.1. Biological oxygen demand, chemical oxygen demand and dissolved oxygen concentrations in surface water bodies in Addis Ababa and Akaki area (EEPA 1997).

	Sample Location	BOD mg/l	COD mg/l	DO mg/l
Little Akaki basin	Tributaries of Little Akaki near Asko Shoe Factory	8	48	6.9
	Little Akaki near ALERT bridge	8	23	5.8
	Little Akaki down stream of Medhane Alem School bridge	301	621	0
	Little Akaki near Mekanissa Bridge	19	56	4.4
	Little Akaki near Mekanissa Liquor Factory bridge	254	375	1.3
	Little Akaki inlet to Aba Samuel Lake	321	708	0
	Outlet from Aba Samuel Lake	0.8	15.6	7.4
Big Akaki Basin	Kebena (French Embassy Bridge)	11	43	6.8
	Kebena at Bole Bridge	29	105	7
	Big Akaki inlet to Aba Samuel Lake	13	28	5.2
	Outlet from Aba Samuel Lake	0.8	15.6	7.4

Sample locations are listed from upstream to downstream part of the Small and Big Akaki Rivers

High concentrations of nitrate are the main problems in most surface water and groundwater samples. Along with some heavy metals and bacteria this high concentrations of nitrate is the main cause of child health problems and deaths in the outskirts of Addis Ababa where the problem is critical.

The chemical, physical and microbiological characteristics of various sources of water in the surrounding of Addis Ababa were also presented (Worku et al. 1998). In this study 18 samples were collected from reservoirs that supply Addis Ababa's tap water (Lakes Gefersa & Legadadi) city taps, city swimming pools, Ambo Mineral Water, rain water, Akaki and Bulbula Rivers and Lake Aba Samuel (Table 2.2).

Table 2.2. Mean concentrations of heavy metals (ppb) and pH in Addis Ababa Catchments (Worku et al. 1998).

Heavy metals	Streams	Springs	Boreholes
pH	7.72	6.61	8.62
Mn	2187	29.8	5.14
Cr	4.24	1.84	1.30
Ni	9.03	0.32	0.51
As	1.2	8.44	0.44
Pb	0.0	4.64	16.6
Zn	0.0	3.05	35.3

The results revealed that samples taken from the Akaki and Bulbula Rivers pose high health risk because of their high concentration of Cr^{3+}, Ni^{2+}, As^{3+}, Pb^{2+}, Mn^{4+} and Zn^{2+}.

Itanna (1998) has shown that the pollution of edible vegetables that were irrigated by wastewater released from industries in and around Addis Ababa endangered health of the communities. In the Akaki area, liquid waste released by the Akaki Textile Factory has been used by the surrounding villagers to produce vegetables and fruits. Itanna (1998) described the concentrations of metals in leafy vegetables and their toxicological implications in Addis Ababa. In this study, he pointed out that the concentrations of As, Cr, Fe and Pb have surpassed maximum permitted concentrations at Kera farms that are located close to the Small Akaki River. The metal concentrations of some vegetables irrigated with industrial liquid waste at the Akaki area has also been investigated (Itanna 1998). In this study, vegetables such as Swiss chard (Beta vulgaris var cicla L.), onion (Allium cepa L.), potato (Solanum tuberosum L.), cabbage (Brassica oleracea L. var capitata), and red beet (Beta vulgaris L.) were tested for their contents of heavy metals that have toxicological significance. These vegetables are irrigated with industrial liquid waste that is discharged directly by the Akaki Textile Factory. The concentrations of Cd, Cr, Cu, Hg, Ni and Zn of potato have surpassed the maximum limits expected at background concentration levels (Itanna 1998). Chromium concentrations were also above the background values in onion and red beet while Hg concentrations in all vegetables were above the background (Table 2.3). The concentrations of Cd, Cr, Cu, Hg, Ni, and Zn in potato have exceeded the expected concentration levels set by Weigert (1991) for food crops. Swiss chard, cabbage, red beet, and onion irrigated with the same waste water showed elevated concentrations of As, Cu, Cd, Cr, Ni, and Zn than the mean value stated for vegetables (Weigert 1991). The contents of Hg in all vegetables surpassed the maximum tolerable level found in foodstuffs.

Table 2.3. Leaf concentrations of arsenic and some heavy metals of different vegetables at maturity stage, grown with industrial liquid waste at Akaki (Ittana 1998).

Vegetables	Metal content (mg/kg)						
	As	Cd	Cr	Cu	Hg	Ni	Zn
Cabbage	0.105	0.030	1.80	3.28	0.218	0.64	29.7
Onion	0.105	0.108	2.81	5.24	0.201	0.44	15.4
Potato	0.133	0.076	2.26	8.72	0.355	1.75	47.4
Red Beet	0.170	0.057	2.87	8.92	0.142	1.47	27.1
Swiss chard	0.038	0.044	1.25	8.96	0.218	0.79	38.1

The concentrations of total Cr, and Ni in the contaminated soils of Akaki reached toxic levels according to the critical limits set by Hein and Schwedt (1991), (Table 2.4). In support of this study, the concentrations of total Cr and Ni in the uncontaminated Akaki soils indicated that these soils are also in the toxic range. This implies that the geogenic sources of contamination also played major roles in the pollution problem.

In his report, Itanna (1998) also mentioned that the Akaki area soils have also been polluted by wastewater that drains the nearby agricultural fields. The total metal concentrations in treated and untreated Akaki soils are presented in table 2.4. Soils that have been treated with the wastewater were enriched in heavy metals as compared to the natural untreated soils.

Table 2.4. Total metal concentrations of treated and untreated soils at Akaki (Itanna 1998).

Sample	Description	Metal content (mg/kg)								
		As	Cd	Co	Cr	Cu	Hg	Ni	Pb	Zn
1	Untreated	3.62	0.10	26.4	82.0	31.2	0.02	46.41	9.11	115
2	Treated	3.82	0.19	35.3	109	40.7	0.08	69.61	9.10	132
3	Treated	3.55	0.18	34.6	112	39.9	0.06	62.20	8.43	142
4	Treated	4.30	0.17	44.4	116	34.8	0.08	53.23	13.06	137
5	Treated	4.41	0.16	45.0	81	33.0	0.09	48.74	17.02	153
Toxic limits**		>20	>3	>50	>100	>100	>2	>50	>100	>300

*Treated samples are taken at 25 m interval down slope.
**According to Hein and Schwedt (1991) and Pendias and pendias (1984).

The dominant types of soils the vegetables are grown on are Pellic Vertisol and Eutric Fluvisol (Itanna et. al 2003). These soils have pH values ranging from 7.1 to 8.3. At the river banks of the Big and Small Akaki Rivers, irrigation of agricultural fields with the river waters are one of the most common practice. The vegetables grown this way constitute the biggest proportion in the diets of the people of Addis Ababa.

3. Behavior of Heavy metals in Terrestrial and Aquatic Environments

3.1. Heavy Metal Pollution and Toxicity

Heavy metals are those metals (i) with atomic number 23 (i.e. vanadium) onwards except Rb, Sr, Y, Cs, Ba, and Fr; (ii) metals with density greater than 5; and (iii) metals that are toxic to man and other life forms when found in fairly low concentrations. Copper, Zn, Ni, As, and Pb are among the eight most common heavy-metal/metalloid pollutants listed by the Environment Protection Agency (EPA). Rapid developments and increase in mining and industrial activities have gradually redistributed many of the toxic metals from the earth's crust to the environment. This has in turn raised the chances of human exposure to these metals (in excess of their natural levels) through ingestion, inhalation or skin contact (Nriagu 1988).

Heavy metals are often problematic environmental pollutants, with well-known toxic effects on living systems. Nevertheless, because of certain useful physical and chemical properties, some heavy metals, including Pb and Cd, are intentionally added to certain consumer and industrial products such as batteries, switches, circuit boards, and certain pigments. Metals are partitioned in the biota by different acid-base affinities, by their kinetics, by spatial partitioning (e.g., by membranes and compartments), and by temporal partitioning (Fraústo da Silva and Williams 1991)

Lead Zn, Cd, Cr, Cu, Ni are also commonly classified as heavy metals, subset of elements that occur in low concentration in the earth's crust (Table 3.1). All soils naturally contain trace levels of metals. The presence of metals in soil is, therefore, not indicative of contamination. The concentration of metals in uncontaminated soil is primarily related to the geology of the parent material from which the soil was formed (McLean 1992). Depending on the local geology, the concentration of metals in soil may exceed the ranges listed in table 3.1. The persistence of heavy metals in soils is a long process (Alloway 1990, Kelly 1996). Heavy metals also include essential elements like Fe, Mn and Zn.

Heavy metals may come from many different sources in urbanized areas, including vehicle emissions, industrial discharges and other activities (Harrison, Gibson and Thornton 1991). The sources of metal contamination of aquatic ecosystems include sewage and industrial discharges, domestic wastewaters, base-metal mining and smelting, atmospheric deposition, etc. Fresh water resources are particularly prone to external trace-metal inputs.

Another sinks for heavy metals are soils. Heavy metals/metalloids being immobile in soils accumulate in the topsoil, thus endangering crops and vegetables. Table 3.2 summarizes the most important heavy metals encountered in natural waters. Some of these are recognized as nutrients required for animal and plant life, including some that are essential at low levels but toxic at higher levels.

Table 3.1. Content of various elements in soils (Modified after Lindsay 1979).

Metal	Selected average contents in soils mg/kg	Common range for soils
Fe	38,000	7,000 - 550,000
Mn	600	20 - 3,000
Cu	30	2 - 100
Cd	0.06	0.01 - 0.07
Zn	50	10 - 300
As	5	1.0 - 50
Ni	40	5 - 500
Pb	10	2 - 200

Aquatic contamination by heavy metals is very harmful since these elements are not degradable in the environment and may accumulate in the living organisms. Heavy metals can combine and divert many aspects of the physiological systems of life (enzymes, chlorophyl operations). Due to their affinity for sulfur (a constituent of some amino acids and, therefore, of protein), they disrupt enzyme activity by forming bonds with sulfur. In addition to this, carboxylic acid groups and amine groups that are present in protein also form strong attachments to metals (Manahan 1996).

Some of the heavy metals are among the most harmful of the elemental pollutants and are of particular concern because of their toxicities to humans (Luoma 1983). These elements are, in general, the transition metals, and some of the representative elements, such as Pb, Cd, Ni, and As (Manahan 1996). Heavy metals that include the essential elements such as Fe, Mn and Zn also have tremendous affinity to sulfur, and disrupt enzyme function by forming bonds with sulfur groups in enzymes. Protein carboxylic acid ($-CO_2H$) and amino ($-NH_2$) groups are also chemically bound by heavy metals. Cadmium, Cu, Pb, and Hg ions bind to cell membranes, hindering transport processes through the cell wall. Heavy metals may also precipitate phosphate bio compounds or catalyze their decomposition (Manahan 1996).

Heavy metals can pose health hazards if their concentrations exceed allowed maximum limits. Even when the concentration of metals does not exceed these limits, there is still a potential for long-term contamination, since heavy metals are known to be accumulative within biological systems. The toxicity of heavy metals is discussed briefly in Whitton and Say (1975), Alabaster and Lioyd (1980), Laws (1981), Leland and Kuwabara (1985), Hellawell (1986), Mance (1987), Furness and Rainbow (1990), Newman and McIntosh (1991), Kennish (1992), and Depledge et al. (1994). The cupric ion (Cu^{2+}) is the most toxic species of Cu to plants and algae when present at higher levels. Copper toxicity has also been demonstrated for $CuOH^+$ and $Cu_2(OH)^{2+}$ (LaGrega et al. 1994). Lead is one of the most ubiquitous toxic metals. Lead blood levels as low as <1 - 100 ug/dl in children, and 10 - 100 ug/dl in adults have been associated with a wide range of adverse effects such as nervous system disorders, anaemia and decreased haemoglobin synthesis, cardiovascular disease, and disorders in bone metabolism, renal function and reproduction. Of particular concern is the effect of relatively low exposure on cognitive and behavioral development in children (Pirkle et al. 1998, USPHS 1997, Bernard et al. 1995, Goyer 1993, Nriagu 1988). Long term exposure to Ni results in serious consequences such as chronic bronchitis and reduced lung function (USPHS 1997). Pregnancy complications in Ni-exposed workers including cardiovascular and musculoskeletal defects have also been reported (Chashschin et al. 1994). Excessive dietary exposure of Zn in both humans and animals can cause gastrointestinal distress and diarrhea, pancreatic damage and anaemia (USPHS

1997, Goyer 1996). Arsenic (e.g., As^{3+}) can be toxic through its interaction with sulfhydryl groups of proteins and enzymes (to denature the proteins and enzymes within the cells; Gebel 2000, Graeme & Pollack 1988) and through an increase of reactive oxygen species in the cells, consequently causing cell damage (Ahmed et al. 2000, Wang et al. 1996, Chen et al. 1998 & Nies 1999)

Table 3.2. Source and behavior of heavy metals and arsenic (Adriano 2001).

Element	Sources	Effects and Toxicity
Arsenic	Mining by-product, chemical waste	Toxic, possibly carcinogenic
Copper	Metal Plating, mining, industrial waste	Essential trace element, toxic to plants and algae at higher levels
Iron	Industrial wastes, corrosion, acid mine water, microbial action	Essential nutrient, damages fixtures by staining
Lead	Industrial waste, mining, fuels	Toxic, harmful to wildlife
Manganese	Industrial wastes, corrosion, acid mine water, microbial action	Toxic to plants, damages fixtures by staining
Zinc	Industrial waste, metal plating, plumbing	Essential element, toxic to plants at higher levels

3.2 Chemical Fate and Mobility of Heavy Metals

Metals can be found in the three-phase system: solid, gaseous, or liquid phases in natural soils, sludges, and solid wastes. The fate and transport of a metal in soil and groundwater depends significantly on the chemical species and speciation of the metal (Allen et al. 1991). In surface-and ground-water systems the mobility of metals is hindered by reactions that cause metals to adsorb or precipitate, or chemistry that tends to keep metals associated with the solid phase and prevent them from dissolving. These mechanisms can retard the movement of metals and also provide a long-term source of metal contaminants (NRC 1994). While the various metals undergo similar reactions in a number of aspects, the extent and nature of these reactions vary under particular conditions. The extent of sorption of several metal cations and anions onto iron oxide is shown as a function of pH for a particular background electrolyte composition (Fig. 3.1).

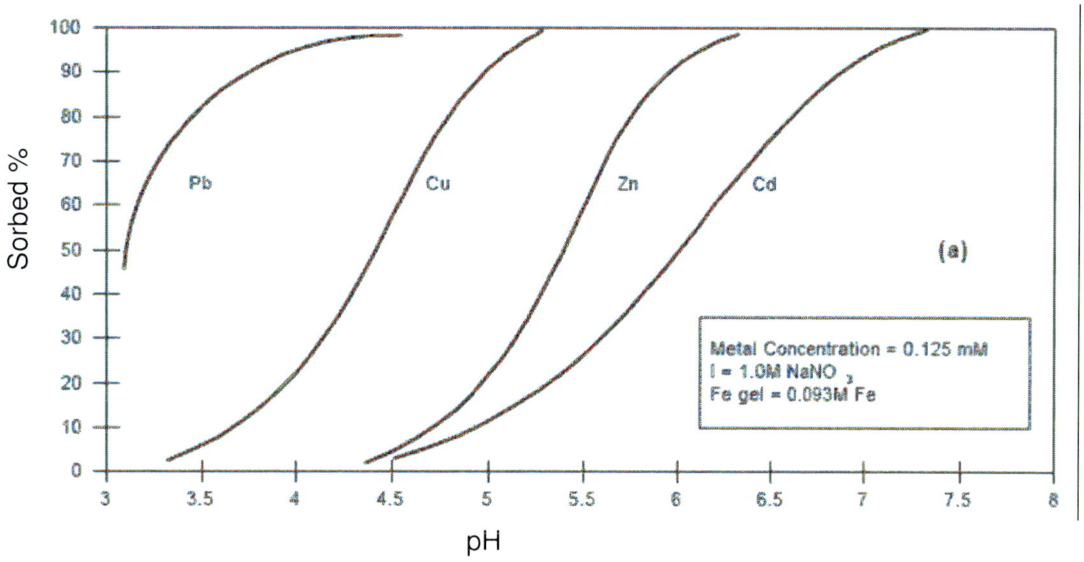

Fig. 3.1. Heavy metal adsorption to hydrous iron oxide gels (Kinniburgh et al. 1976).

It may be seen that Pb sorbs extensively than Zn or Cd to the iron oxide gels at much lower pH values (Kinniburgh et al. 1976). In figure 3.2, the extent of sorption of metal anions onto iron oxide is also shown as a function of pH for a particular background electrolyte composition. The chemical species and speciation of some of the most important metals such as Pb, As, Zn, Ni and Cu are discussed below. The influence of chemical species on fate and mobility of these compounds is also discussed.

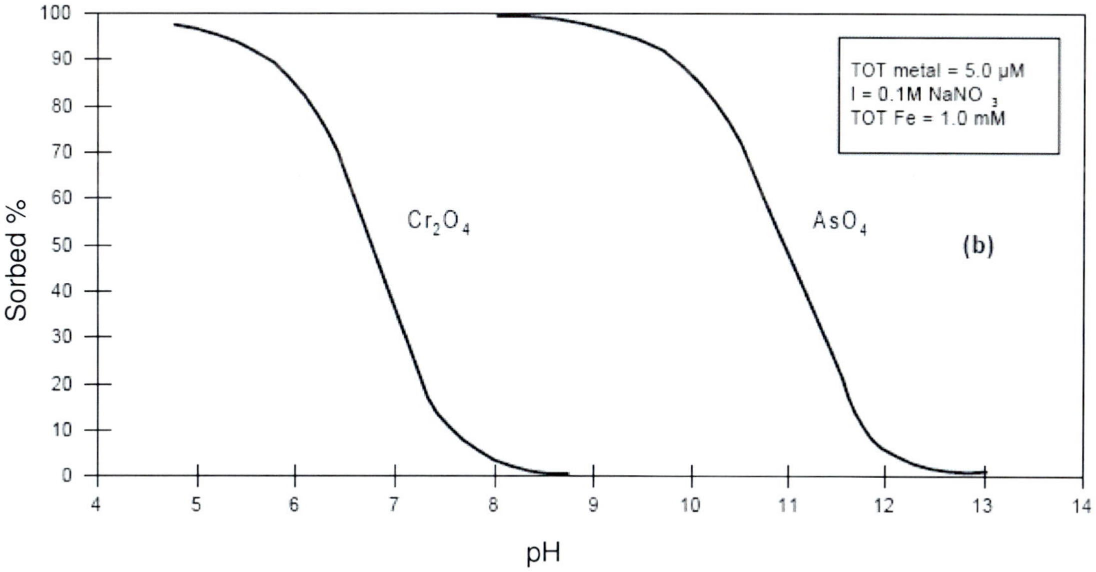

Fig. 3.2. Metal anions adsorption to hydrous iron oxide gels (Leckie et al. 1980; Honeyman et al. 1984).

Lead occurs naturally in the environment. However, most Pb concentrations that are found in the environment are a result of human activities. Due to the application of lead in gasoline an unnatural lead-cycle has consisted. The primary industrial sources of Pb contamination include metal smelting and processing, secondary metals production, lead battery manufacturing, pigment and chemical manufacturing, and lead-contaminated wastes (Evanko and Dzombak 1997). There are different forms in which Pb is released in to groundwater, surface water and soil. These are elemental Pb, Pb oxides and hydroxides, and Pb metal oxyanion complexes (Smith et al. 1995). Pb^0 or Pb^{2+} is the most commonly known oxidation states of lead. However, Pb^{2+} is the more common and reactive form of Pb and forms mononuclear and polynuclear oxides and hydroxides. Under most environmental conditions Pb^{2+} and Pb-hydroxy complexes are the most stable forms of lead (Smith et al. 1995). Lead compounds formed by complexation reactions with inorganic (Cl^-, CO_3^{2-}, SO_4^{2-}, PO_4^{3-}) and organic ligands (humic and fulvic acids, EDTA, amino acids) are less soluble (Bodek et al. 1988). When high sulfide concentrations are present under reducing conditions the most stable form of Pb is PbS while Pb carbonate solids form when the pH is above 6 (Fig. 3.3). Most Pb that is released to the environment is retained in the soil (Evans 1989). The primary processes influencing the fate of Pb in soil include adsorption, ion exchange, precipitation, and complexation with sorbed organic matter. These processes limit the amount of lead that can be transported into the surface water or groundwater. The relatively volatile organolead compound tetramethyl lead may form in anaerobic sediments as a result of alkyllation by microorganisms (Smith et al. 1995).

The amount of dissolved Pb in surface water and groundwater depends on pH, the concentration of dissolved salts and the types of mineral surfaces present. In surface water and groundwater systems, a significant fraction of Pb is undissolved forming ligand complexes and occurs as precipitates ($PbCO_3$, Pb_2O, $Pb(OH)_2$, $PbSO_4$), as sorbed ions or as surface coatings on minerals, or as suspended organic matter (Fig. 3.3).

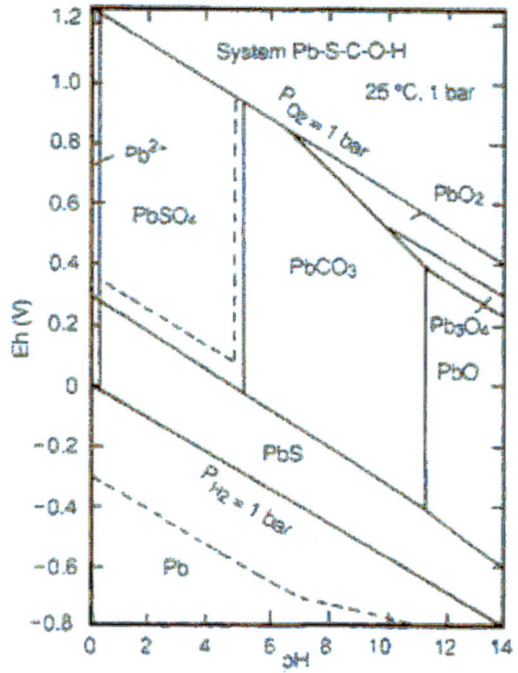

Fig. 3.3. Eh-pH diagram for part of the Pb-S-C-O-H system (Brookings 1988). Assumed activities of dissolved species: Pb=10^{-6}; S=10^{-3}; C=10^{-3}.

Arsenic (As) is a semi metallic element that occurs in a wide variety of minerals, mainly as As_2O_3, FeAsS and can be recovered from processing of sulfidic ores containing mostly Cu, Pb, Zn, Ag and Au. It is also present in ashes from coal combustion (Evanko and Dzombak 1997). The most common oxidation states of arsenic are (As^{3-}, As^0, As^{3+}, As^{+5}) (Smith et al. 1995) and thus exhibit fairly complex chemistry. In aerobic environments, As(V) is the dominant species, usually in the form of arsenate (AsO_4^{3-}) in various protonation states: H_3AsO_4, $H_2AsO_4^-$, $HAsO_4^{2-}$, AsO_4^{3-}. Arsenate and other anionic forms of As behave as chelates and can precipitate when metal cations are present (Bodek et al. 1988). Metal arsenate complexes are stable only under certain conditions. Under acidic and moderately reducing conditions As(V) can also coprecipitate with or adsorb onto iron oxyhydroxides. Coprecipitates are immobile under these conditions but As mobility increases as pH increases (Smith et al. 1995).

Arsenite (AsO_3^{3-}) and its protonated forms: H_3AsO_3, $H_2AsO_3^-$, $HAsO_3^{2-}$ are the dominant forms under slightly reducing conditions (Fig. 3.4). Arsenite can adsorb or coprecipitate with metal sulfides and has a high affinity for other sulfur compounds.

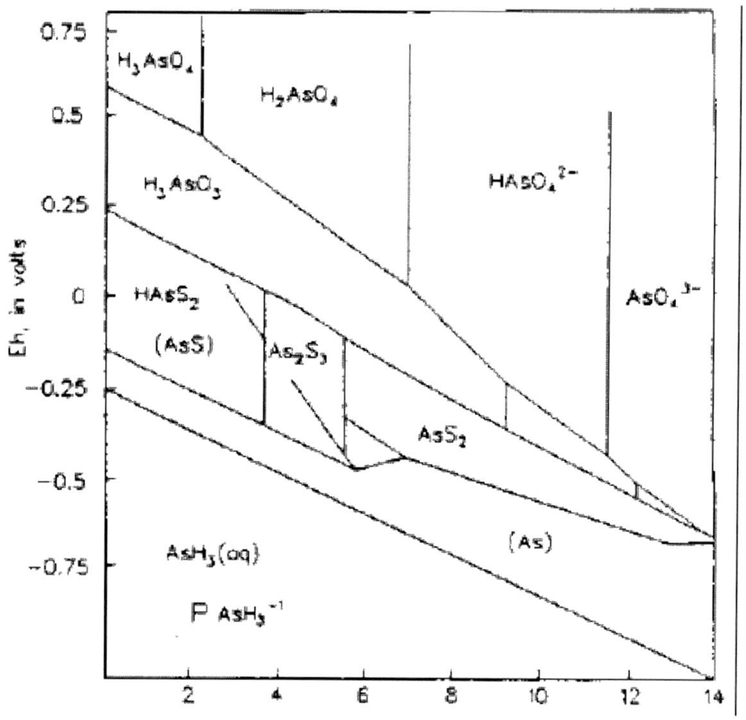

Fig. 3.4. Eh-pH diagram of the system As-H_2O-S at 25 °C (Fergusson 1990).

The elemental As and arsine, AsH_3, may be present under extreme reducing conditions. Biotransformation (via methylation) of As creates methylated derivatives of arsine, such as dimethyl arsine $HAs(CH_3)_2$ and trimethylarsine $As(CH_3)_3$, which are highly volatile.

Since As is often present in anionic form, it does not form complexes with simple anions such as Cl^- and SO_4^{2-}. Arsenic speciation also includes organometallic forms such as methylarsinic acid $(CH_3)AsO_2H_2$ and dimethylarsinic acid $(CH_3)_2AsO_2H$.

Many As compounds sorb strongly to soils and are therefore transported only over short distances in groundwater and surface water. Sorption and coprecipitation with hydrous iron oxides are the most important removal mechanisms under most environmental conditions (Krause and Ettel 1989, Pierce and Moore 1982). Oscarson et al. (1981) hypothesized that the oxides of Mn(IV) and Fe(III) may be the primary electron acceptors in the oxidation of As(III) because Mn and Fe readily participate in many oxidation-reduction reactions in natural environments. Arsenates can be leached easily if the amount of reactive metal in the soil is low. As(V) can also be mobilized under reducing conditions that encourage the formation of As (III), under alkaline and saline conditions, in the presence of other ions that compete for sorption sites, and in the presence of organic compounds that form complexes with As (Smith et al. 1995).

Zinc (Zn) does not occur naturally in elemental form. It is usually extracted from mineral ores to form Zn oxide (ZnO, $Zn(OH)_2$) and $ZnCO_3$. The primary industrial use for Zn is as a corrosion-resistant coating for iron or steel (Smith et al. 1995).

Zinc usually occurs in the Zn^{2+} oxidation state and forms complexes with a number of anions, amino acids and organic acids. The most common Zn precipitates are Zn $(OH)_{2(s)}$, $ZnCO_{3(s)}$, and $ZnS_{(s)}$.

Since Zn is present as soluble compounds at neutral and acidic pH values, it is one of the most mobile heavy metals in surface waters and groundwater (Smith et al. 1995). At higher pH values, Zn can form carbonate and hydroxide complexes that control Zn solubility (Fig. 3.5). Zinc readily precipitates under reducing conditions and in highly polluted systems when it is present at very high concentration, and may coprecipitate with hydrous oxides of Fe or Mn (Smith et al. 1995). ZnS is another form of Zn precipitate. Due its amphoteric properties, Zn forms variety of salts. Zinc chlorate, Zn chloride, the sulfates and nitrates are readily soluble in water. The oxide, carbonate, phosphates, silicates, sulfides and organic complexes of Zn are practically insoluble in water (Weast 1986).

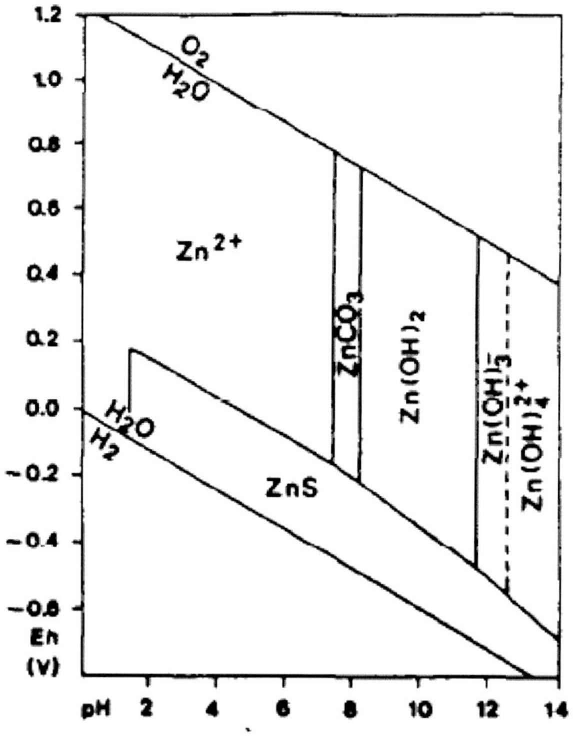

Fig. 3.5. Eh-pH diagram for Zn-CO_2-S-H_2O system at 25℃ and 1 atm (Wyman 1993).

The primary fate of Zn in aquatic environments is mainly controlled by sorption to sediments or suspended solids, including hydrous Fe and Mn oxides, clay minerals, and organic matter.

Compared to other inorganic contaminants of environmental interest, the geochemical behavior of Ni in soil and groundwater systems has not been studied extensively (Evanko and Dzombak 1997). However, some works of Rai et al. (1984) present a detailed review of the aqueous speciation, solubility, and adsorption reactions that affect the environmental mobility of Ni. In aqueous systems, the most important oxidation state of Ni is Ni^{2+} (Baes and Mesmer 1976). At pH values less than 10, the uncomplexed cation Ni^{2+} is predicted to be the dominant aqueous species of Ni. At pH values greater than 10, dissolved Ni is present as a series of hydroxide species (Fig. 3.6). To a limited degree Ni also forms aqueous complexes with ligands, such as dissolved sulfate, carbonate, and chloride (Rai et al. 1984).

Under reducing geochemical conditions, Ni may potentially precipitate in soils and control the maximum concentration of total dissolved Ni. The results of a solubility study by Mattigod et al. (1997) indicate that solid $Ni(OH)_2$ (the mineral theophrastite) may precipitate and control the maximum concentration of dissolved Ni in soils under alkaline conditions. It was indicated that at concentrations of dissolved Ni less than 10^{-7} mol/l bunsenite (NiO) is oversaturated at basic pH values.

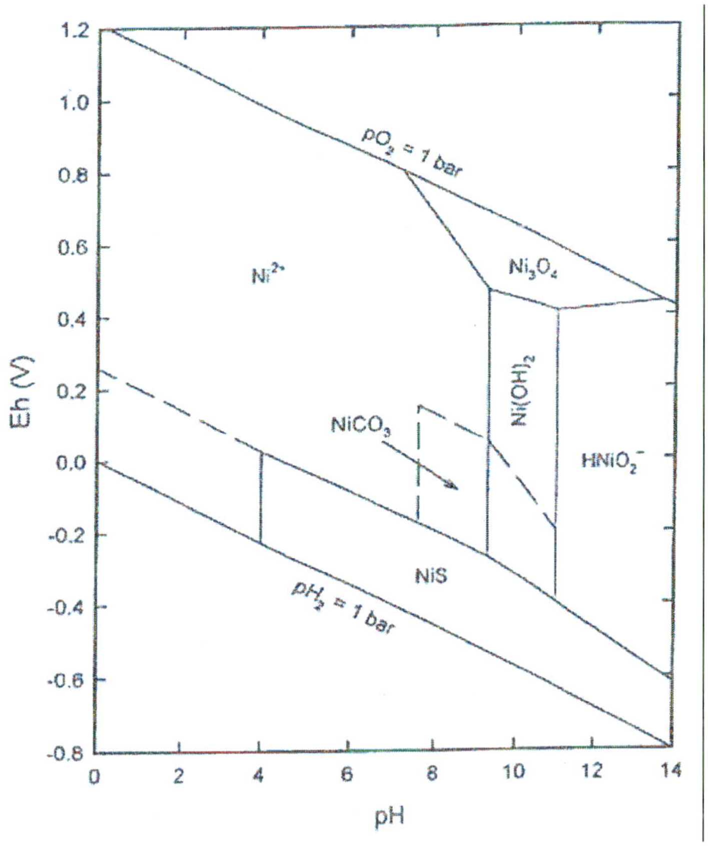

Fig. 3.6. Eh-pH diagram for part of the system Ni-O-H-S-C. Assumed activities for dissolved species are: Ni=10^{-4}, S=10^{-3}, C=10^{-3} (all in mol/l), at 298 K and 1 bar (Bermanec et al. 2000).

Surface sorption processes likely control the concentrations of Ni in most soil systems. Nickel is known to be adsorbed by Fe and Mn oxides and clays (Rai et al. 1984 and references therein). The adsorption of the uncomplexed cation Ni^{2+} will be greatest at high pH values less than 11, decrease with decreasing pH, and be minimal at acidic conditions. At pH values greater than 11, the adsorption of Ni to soil may decrease if the dominant aqueous species is anionic, such as $Ni(OH)^{3-}$ (Fig. 3.6). Recent studies by Nachtegaal and Sparks (2003), Elzinga and Sparks (1999, 2001), Scheckel and Sparks (2000, 2001), and other related papers by these authors demonstrate that the sorption of Ni to some clay minerals results in the formation of Ni hydroxide or Ni aluminum hydroxide surface precipitates on clays. The formation of such surface precipitates will reduce the migration and remobilization of Ni in soil-water systems.

Copper (Cu) is mined as a primary ore product from Cu sulfide and oxide ores. Mining activities are the major source of Cu contamination in groundwater and surface waters. Other sources of Cu include algaecides, chromated copper arsenate (CCA) pressure treated lumber, and Cu pipes. Anthropogenic releases of Cu from industrial wastes are also potential Cu sources in the environment (Evanko and Dzombak 1997).

In ground-water systems, solution and soil chemistry strongly influence the speciation of Cu. $CuCO_3$ is the dominant soluble Cu species in aerobic, sufficiently alkaline systems. The cupric ion, Cu^{2+}, and hydroxide complexes, $CuOH^+$ and $Cu(OH)_2$, are also commonly present (Fig. 3.7). Copper forms strong solution complexes with humic acids. The affinity of Cu for humates increases as pH increases and ionic strength decreases.

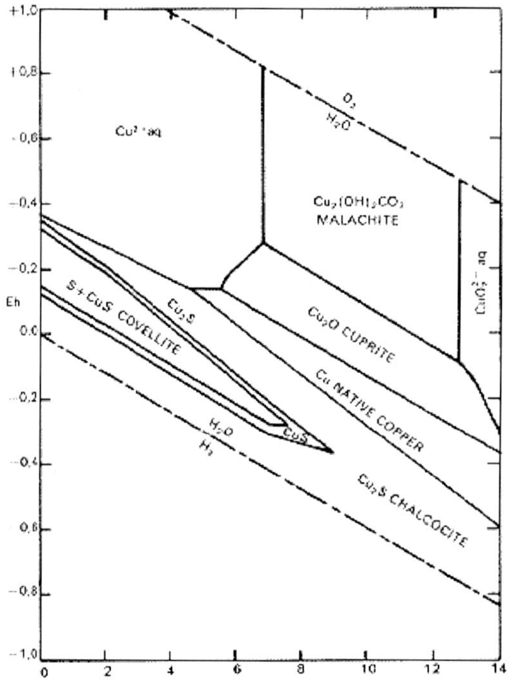

Fig. 3.7. Eh-ph diagram for $Cu\text{-}H_2O\text{-}O_2\text{-}S\text{-}CO_2$ System (Garrels and Christ 1965).

In anaerobic environments, in the presence of sulfur, the formation of CuS(s) is highly favored. The mobility of Cu in the environment is decreased by sorption to mineral surfaces. Cu^{2+} sorbs strongly to mineral surfaces over a wide range of pH values (Dzombak and Morel 1990). The cupric ion (Cu^{2+}) is the most toxic species of Cu. The importance of redox potential in determining metal solubility and transport can be clearly seen for Cu in the Eh-pH diagram for the Cu-H_2O-O_2-S-CO_2 system (Fig. 3.7, Garrels and Christ 1965).

3.3 Treatment of Heavy Metals Contaminated Water

Many researchers have investigated thc rcmoval of heavy metals from contaminated industrial wastes using economically feasible materials (Deans and Dixon 1992). The direct release of these wastewaters in to the environment may result in the contamination of various portions of the environment. Since industrial wastes contain in many instances toxic heavy metals there is high probability they will incorporate in the important environmental portions such as soil, water and air.

There are various dynamic transformations that may occur when a heavy metal contaminated water passed through a filtration pond (natural media) (Leeuwang 1990 and Johnston 1993). Different transport processes may happen from one compartment to another, e.g. from water to sediments or biota or suspended solids or vice versa. In calm waters particles that are denser than water will settle down. The formation of flocs should be effected for particles, which are light or less dense than water, for sedimentation processes to become possible. Particles of clay and organic mater that have surface electronic charge aggregate to form flocs and generally settle more rapidly than do individual particles (Hakanson and Jansson 1983). Flocs may also adsorb other types of suspended particles including heavy metals. Increased pH, turbulence, and concentration of suspended matters, ionic strength and high algal concentration enhance flocculation. Because of their larger surface area they have proportionally greater adsorption potential. The hydrous oxides of Fe and Al carry a positive electrical charge necessary to neutralize the negative charges of colloidal particles resulting aggregation and sedimentation. Sedimentation is a physical process that takes places after other mechanisms aggregate heavy metals into particles large enough to sink. In this way heavy metals are removed from wastewater and trapped in filter media (or in filtration pond).

In sediments heavy metals are adsorbed to clay and organic matter by electrostatic attraction (Patrick et al. 1990). Unlike organic pollutants, which will ultimately decompose, heavy metals will remain as metal atoms once adsorbed on to humic or clay colloids. As the organic molecules decompose or as sediment conditions change metal speciation may change with time. There are many factors that affect the extent to which metals ions are adsorbed by cation exchange or non-specific adsorption. These include properties of the metals concerned (valence, radius, degree of hydration and co-ordination with oxygen). Other factors are physico-chemical environment (pH and redox conditions), the nature of the adsorbent medium (permanent and pH-dependent charge, complex-forming ligands), and the concentrations and properties of other metals and soluble ligands present (Alloway 1990). For instance, more than 50% of heavy metals can easily be adsorbed onto particulate matter in wetland and thus be removed from the water component by sedimentation (Muller 1988). The adsorption of metal ions on to mineral surfaces is mostly described either by the Langmuir or the Freundlich adsorption isotherms. Metal adsorption onto Mn oxide, for instance, can be described by the Langmuir equation for a range of metal concentrations, over about one order of magnitude only (Van den Berg 1982).

Another important mechanism in contaminated water treatment is co-precipitation of heavy metals with secondary minerals, including the hydrous oxides of Fe, Al and Mn. Heavy metals such as Cu, Mn, Ni, and Zn are coprecipitated in Fe oxides while Fe, Ni, Pb and Zn are co-precipitated in Mn oxides. The precipitation of Fe^{3+} may occur initially in the form of gelatinous stable forms, such as goethite. Through the decrease in Eh or pH ferrihydrite is more likely to be subsequently dissolved than goethite again. Ferrohydrite co-precipitates other ions and as a result of its large surface area acts as a scavenger sorbing both cations, such as heavy metals and anions, especially HPO_4^{2+} or $H_2PO_4^{+}$ and $H_2AsO_4^{-}$. In reducing conditions pyrite (FeS_2) forms when sulfate becomes reduced to S^{2-}, producing H_2S which then reacts with Fe^{2+} to form FeS and FeS_2. The oxidation of S^{2-} such as pyrite causes marked acidification in soils and sediments. This causes heavy metals to be released into solution. Specialized bacteria, e.g. Thiobacillus ferroxidans and Metallogenum spp are involved in the transformations of Fe and Mn respectively. Fe and Mn oxides occur as coatings on soil particles, fillings in voids and as concentric nodules. The oxide coatings are normally intimately mixed with clay and humus colloids and, although mineralogically distinct, form part of the clay-sized fraction. The heavy metals normally found co-precipitated with secondary minerals in soil sediments are (Sposito 1983):

Fe oxides:	Mn, Ni, Cu, Zn
Mn oxides:	Fe, Ni, Zn, Pb
Ca carbonates:	Fe, Ni, Cd
Clay minerals:	Ni, Zn, Pb, Mn, Fe

When reducing conditions cause the dissolution of hydrous Mn and Fe oxides, the concentrations of several other elements in the sediment solution are likely to increase. In this case Cu, Ni, Fe, and Mn are generally more bioavailable from gleyed (periodically water logged soils) than from drained wetlands soils on the same parent material (Matagi et al. 1998). However, Cu and Zn do not undergo redox reactions themselves but are coprecipitated by hydrous oxides.

Precipitation is one of the major mechanisms by which metals are removed from water deposited in sediments. Precipitation is affected by several factors. Solubility product (K_{sp}) of the metal species involved, pH and concentration of metal ions and relevant anions are some of the most important factors that affect precipitation. Under reducing conditions, carbonates, hydroxides and sulfides of metals are precipitated and their precipitation is also pH dependent. More specific for sulfides is that they are insoluble at neutral pH and therefore accumulate in fresh water sediments. For the carbonates, the solubility is also influenced by partial pressure of CO_2. For example the solubility of $PbCO_3$ can be increased several fold in the present of CO_2. The bioavailability of heavy metals to many aquatic ecosystems is mainly controlled by the formation of insoluble heavy metal precipitates.

Ion exchange is another removal mechanism that takes place in water treatment. It can occur between the counter ions balancing the surface charge on the sediment colloids and the ions in solution. Negative charges on mineral surfaces are responsible for cation exchange, in which exchange of a hydrogen ion for the metal occurs. The extent to which the sediment constituents can act as cation exchangers is expressed as the cation exchange capacity (CEC), measured in meq/100 g. The negative charges on the surface of minerals are of two types:

(a) Isomorphic substitution (permanent charge): These are results of the isomorphous substitution of a clay mineral constituent by an ion with a lower valence. Isomorphic substitution is the replacement of one atom by another of similar size in a crystal lattice without disrupting or changing the crystal structure of the mineral. In many soils the solid phase carries a net negative surface charge. For clay minerals this charge is a result of isomorphous substitution, where structural cations of higher valence are replaced by cations of lower valence (e.g. Si^{4+} is replaced by Al^{3+}). As a result, clay minerals have a permanent negative surface charge (Sposito 1984).

(b) pH dependent charge occurs on the edges of layered silicates, on variable charged minerals such as oxides of Fe and Al, and organic matter. It is called pH dependent charge because it increases in magnitude as the pH of the aqueous soil environment increases. The pH-dependent charges on the oxides of Fe, Al, Mn, Si and organic colloids are positive at pH below their iso-electric points and negative above their iso-electric points. Under most conditions hydrous Fe and Al oxides have relatively high iso-electric points (> pH 8) and tend to be positively charged whereas clay and organic colloids are predominantly negatively charged under alkaline conditions (Alloway et al. 1997). With most colloids, increasing soil pH, at least up to neutrality tends to increase their CEC. Due to the dissociation of protons from carboxyl and phenolic groups humic polymers in the sediment organic matter fraction become negatively charged. The concept of cation exchange implies that ions will be exchanged between e.g. mineral surface (double diffuse layer) and the surrounding water (Alloway et. al 1997). The relative replacing power of anion on the cation exchange complex will depend on its valence, diameter in the hydrated form and the type and concentration of other ions present in water with the exception of H^+, which behaves like a trivalent ion, the higher the valence, the greater the degree of ion exchange. Ions with a larger hydrated radius have a lower replacing power than ions with smaller radii. For example K^+ and Na^+ have the same valence but K^+ will replace Na^+ owing to the greater hydrated size of the Na^+ ion.

In soils the overall electroneutrality is maintained by an excess of electrostatically attracted counterions in proximity to the charged surface. In the case of negatively charged surfaces a diffuse double layer will result where cations are in excess of anions. For individual sediment constituents, the order of replacement of the heavy metals is (Alloway 1990):

Montmorillonite clay: Pb > Cu > Cd > Zn
Ferrihydrite: Pb > Cu > Zn > Ni > Cd
Peat: Pb > Cu > Cd >= Zn

Clay minerals in general have higher CEC values whereby adsorptions of heavy metals are facilitated (Table 3.3).

Table 3.3. Cation exchange capacities (CEC) for several clay minerals (Grim 1968).

Mineral	CEC (meq/100 g)
Chlorite	10 - 40
Halloysite.$2H_2O$	5 - 10
Halloysite.$4H_2O$	40 - 50
Illite	10 - 40
Kaolinite	3 - 15
Sepiolite-Attapulgite-Palygorskite	3 - 15
Smectite	80 - 150
Vermiculite	100 - 150
Organic matter	50 - 100

The redox state of a heavy metal in solution is an important speciation parameter because it can drastically affect its toxicity, adsorptive behavior and metal transport (Mertz and Cornazer 1971, Florence et al 1983). The redox state of heavy metals varies depending on whether there are anoxic or oxic conditions in aquatic environments. For instance, the oxidation of sulfides is catalyzed by microorganisms, such as *Thiobacillus* spp in water. In the case of pollution by tailings from metalliferous mining, particles of ore minerals in soils, such as, PbS, ZnS and $CuFeS_2$ become oxidized, releasing metal cations Pb^{2+}, Cu^{2+}, Zn^{2+} and Cd^{2+} into the sediment when they are adsorbed. Charges in redox potential Eh under reducing conditions allows the metals to precipitate as metal sulfides.

4 Methods and Materials

4.1 Logistics and Planning

The main objective of this PhD work is to propose a concept to purify contaminated water in the Akaki area that is used for drinking and other household purposes. The need to conduct sampling in the area was necessary. In order to carry out sampling of different water sources and potential filter media around the Akaki area preliminary information have been gathered from different governmental and non-governmental institutions. The institutions were also co-operative to provide such information as previous data on contamination analysis, geological and topographic maps of Addis Ababa and the surrounding area. Geological maps (1:50,000) of Addis Ababa, the Akaki area, and the southern and southwestern peripheries of Addis Ababa were provided by the Ethiopian Mapping Authority (reprinted August 2000). The maps were used as basis for traversing the target areas.

To assist our sampling campaign and research work mutual collaborations have been established between the University of Karlsruhe, Institute of Mineralogy and Geochemistry (IMG) and, Addis Ababa University, Department of Geology and Geophysics. In this regard, a co-operation memorandum has been signed between the two parties to carry out mutual research activities, technology transfer and exchanging graduate and postgraduate students between them.

4.2 Sampling Procedure

Between April and May 2003, a field trip to Ethiopia had been employed to carry out sampling and some field measurements in the study area. The samples were collected during the dry season in a short tropical monsoon rainy season. This period represents end of the dry season and beginning of short rainy season in the region. The sampling campaigns included (I) collection of water samples from groundwater, rivers, lakes, ponds and industrial wastewater and (II) gathering of natural filtering media from the study area and the surrounding. The sampling of filter media has also been employed from the areas located between the Addis Ababa - Awassa town. Whenever possible, the potential targets to collect filter materials were selected as close to the contaminated water resources as possible. This enables the local communities to access the filter materials without further transport cost. On the other hand, water samples representing the wet season pollution level of the Akaki waters were collected between September and October 2005.

4.2.1 Water Samples

During the April-May, 2003 sampling campaign 22 water samples out of which 6 surface water samples, 15 groundwater samples and 1 industrial wastewater sample were collected from the Akaki Sub-Basin (Table 4.1). Samples collected during the wet season (October 2005) included 10 groundwater and 2 surface water samples (appendix I and II). All water samples were collected by grab method from the southern portion of the Akaki area near the Aba Samuel Lake.

Water samples were also collected from the two river systems draining the city of Addis Ababa: Big Akaki and Little Akaki Rivers at different sampling points. Groundwater wells close to the Akaki Rivers and Aba Samuel Lake were also sampled. These water wells were basically drilled very close to the river systems to get maximum yield not taking much attention to the possible contamination. Samples were treated in the field and at laboratories of Chemistry Department, Addis Ababa University.

4.2.1.1 Sample Preparation

The sample equipment and containers used during sampling are made up of polyethylene. During field water parameter measurements and sampling of the surface and groundwater samples proper preservation and storage were applied for minimizing the irreversible changes that may happen due to the reaction with atmospheric oxygen. Field measurements of the following parameters have been made at the time of sample collection on a portion of each water sample, and recorded in field logbook: pH, temperature, specific conductivity, and dissolved oxygen.

Table 4.1. The types of water samples, sampling sites, and water use (Tolera 2003).

Sample Code	Sample sites	Location			Sample Type	Utilization
		N	E	m		
Ku1	Small Akaki	08°51.080	38°46.906	2062	River water	Irrigation, Drinking
De2	Dewera Guda	08°48.415	38°44.625	2075	Well water	Drinking
Dk3	Dika	08°46.991	38°41.840	2014	Well water	Drinking
Dk4	Aba Samuel	08°47.128	38°42.094	2005	Lake water	Irrigation, Household
Dk5	Aba Samuel	08°47.255	38°42.505	2016	Lake Outlet	Irrigation
De6	Dewera Guda	08°48.415	38°44.625	2075	Well water	Drinking
Gr7	Gerado	08°47.523	38°46.947	2172	Pond water	Drinking, Household
Be8	Berehi	08°94.124	38°48.527	2076	Well water	Drinking
Sd9	Sidamo Awash	08°50.493	38°49.032	2072	Well water	Drinking
E10	EEPCO	08°53.145	38°45.441	2114	Well water	Drinking, Factory
Ju11	Jugan B. Akaki	08°53.181	38°44.823	2076	River water	Irrigation
Tl12	Tilinto	08°53.523	38°47.903	2147	Well water	Drinking
Iv13	Indian Village	08°52.860	38°47.175	1999	Well water	Drinking
Ak14	Akaki Textile F	08°52.674	38°47.064	2010	Effluent	Waste water
Ah15	Aba Hana	08°52.579	38°47.049	2015	Well water	Drinking
Ls16	Lesperance	08°53.127	38°47.633	2089	Well water	Drinking
Ak17	Akaki Metal F	08°51.571	38°47.774	2027	Well water	Drinking, Factory
El18	ELICO	08°56.543	38°45.651	2162	River water	Factory
El19	ELICO	08°56.543	38°45.651	2162	Effluent	Irrigation
Td20	Tulu Dimtu	08°50.907	38°48.503	2035	Well water	Drinking
Am21	Akaki Merenu	08°49.146	38°48.521	2067	Well water	Drinking
Ab22	Akaki Babu	08°50.070	38°48.159	2013	Well water	Drinking

Dissolved oxygen was measured by the WTW Oxi sensor calibrated by an air-calibration chamber that permits calibration of the DO sensor at the temperature of the water in which DO concentration is to be measured. During the calibration procedure, the interior of the chamber was kept just moist. The instrument system used to measure DO was tested before each field trip and cleaned soon after each use.

The electrical conductivity of the water samples were measured in-situ with the WTW Conductivity meter. Calibration of the sensor was performed before the water-quality field trip and checked several times. Surface water conductivity has been measured in situ, whenever possible; otherwise, the determination of conductivity in discrete samples collected from a sampling beaker. The conductivities of samples were measured as soon as possible after collection of the samples. The reported values were determined on unfiltered samples.

pH measurements were performed directly on surface water bodies while groundwater samples were determined in a sampling beaker. The instrument system that was used to measure pH was tested before each field trip, and it was cleaned soon after use. During the pH measurement, buffer solutions of pH 4 and pH7 were used for calibration. In order to avoid pH changes that could be caused by different reasons, the measurements of pHs of the water samples were carried out immediately after withdrawal of discrete samples directly from the water sources or sampling device on unfiltered samples.

During the sampling period, three sets of water samples were collected in 50 ml, 50 ml and 20 ml sampling bottles for analysis of major cations and trace elements, alkalinity, and anions measurements respectively. The overall sampling procedures for all sample types are described below.

Samples for the analysis of heavy metals were collected in thoroughly pre-cleaned polyethylene sampling bottles. The PE bottles were first cleaned with a mixture of super pure nitric acid and distilled water. Sampling bottles were kept screw capped in a dry and clean place before samples were poured in to them. Samples were tapped directly from surface waters with a 100 ml beaker and injected through 0.45 µm cellulose nitrate filter papers (Sartorius AG, Germany) in to the 50 ml polyethylene bottles. For further analysis of heavy metals on the suspended particulate matter the filters were also collected on plastic dishes and kept under +4 °C storage at the Department of Chemistry, Addis Ababa University. In order to preserve the samples at pH 2, 0.1 ml of concentrated HNO_3 (suprapur) was injected in all samples. The samples were kept in a cool dry place under +4 °C before they were measured.

The alkalinities of the water samples were quantified on site after collecting water samples. Samples were filtered through 0.45 µm filter membrane before measuring the alkalinity. The method used to determine alkalinity required electrometric titration of a sample with incremental additions of 0.1 M of concentrated HCl (Gran 1952). The titration procedure applied is summarized as follows.

First, the pH system was calibrated accordingly. Representative samples were collected and sub samples were filtered through 0.45 µm membrane filter. Samples were maintained at ambient water temperature until titration. The electrodes, sensors, beaker, stir bar and delivery tube were rinsed with distilled water. 10 ml of filtered sample was transferred into a beaker using pipette. Conductivity electrode and temperature sensor were inserted away from bottom or sides (Gran 1952). The sample was stirred gently in order to avoid splash of solution and vortex. The initial time, pH, sample volume, titrant normality and conductivity reading were recorded. After adding the titrant, the solution was stirred for 15 - 30 seconds and the final pH, temperature, and conductivity were recorded. The determination of alkalinity was finally done with the help of the Gran function method.

Onsite measurements of ammonium, nitrate, and phosphate were carried out at the Department of Chemistry, Addis Ababa University using a photometer. Test reagents of nitrate, ammonium, and sulphate were used to convert the samples into coloured compounds in a specific reaction. The anions that are found in varying concentrations in the water samples react with the corresponding test kit reagents to produce colored products. The principle behind the reagent tests is that the reagents necessary for the colour reaction are combined in the form of liquid concentrates or solid substance mixtures. In these tests, a few drops of the respective reagent concentrate were added to, for example, 5 ml of water sample. The reagent-sample mixture were shaken well and let for 2 - 3 minutes before inserting the cuvette in the photometer. The accuracy of measurements was checked by comparing the measured values with the measuring range specified for each individual anion. Concentrations were read directly from the display of the photometer.

4.2.2 Filter Media

4.2.2.1 Sampling Site

The filter media represent natural materials of different mineralogical and chemical composition gathered from several localities in Addis Ababa. The sample locations were selected in such a way that they were close to the target area. Twelve natural media of different chemical and mineralogical composition and filtering capacity were collected from Ethiopia. Sample locations were chosen close to the study area whenever possible. However, three filter media were selected for in depth study of their capacities in removing heavy metal from polluted water. These filter media were DQ scoria, KVA volcanic ash, and EBF clay/laterite. For the same investigation, Merapi volcanic materials (PYC pyroclastics) collected from Indonesia were also evaluated. Manganese rich sediment (MnR sediment) as Mn-oxide was also taken for metal removal experiments. The latter, which were supplied by the Institute of Mineralogy and Geochemistry, are also available in other parts of Ethiopia. MnR sediment and PYC pyroclastics have also been investigated by other researchers at IMG to enhance the water quality in small villages in Indonesia.

The exact location of the filter materials including their sample codes and some physical description are presented in table 4.2. Samples were collected by grab method in clean plastic bags. Two-three kg of local materials were sampled. Sample codes and date of collection were written on the sampling bags immediately after sample collection. The samples were kept in airtight plastic bags at room temperature until they were further prepared and analyzed.

4.2.2.2 Sample Preparation

Further characterizations of filter materials in terms of permeability, porosity, and water content were performed at IMG laboratory. Determination of water content at saturation was conducted by carrying out gravimetric water content analysis after drying the filter materials in oven at 105 °C for 24 hours. Porosity, bulk density and permeability were then calculated from the water content values. Curves of grain size distribution were constructed to show the percentages of each size fraction in the filter materials. These were shown after constructing the granulometric analysis curve for each filter material.

Table 4.2. Types, sources, and locations of filter materials.

Sample code	Locality	Description	N	E	Elevation (m)
KVA vol ash	Kumo road	Pumice	08°29.839	39°11.051	1634
DQ scoria	Dika	Scoria	08°47.128	38°42.094	2005
EBF laterite	E. B. F.	Clay	09°04.524	38°41.175	2623
MnR sediment	Indonesia	Mn rich sediment			
PYC pyroc	Indonesia	Volcanic ash			
QTZ sand	IMG lab	Quartz sand			

Six samples have been analyzed for grain size distribution according to the prescription DIN 4022 by sieving dry. Sieves of a mesh size of > 1.6, 1.6,1, 0.5, 0.25, 0.125, 0.05 mm were used in the grain size analysis. Prior to sieving the samples were well dried. The hydraulic conductivity (kf-value) was calculated using the empirical formula by Hazen (1892). Empirical equations have been developed based on grain size data to estimate the hydraulic conductivity (k) of granular materials such as granular soils used for earth dams and sand filters. After Hazen, the hydraulic conductivity,

$$k_f = Cd_{10}^2;$$

With: t = temperature [° C], proportional factor C = (0.7 + 0.03 * t) / 86.4 = 0.0116 at t = 10℃ (~ groundwater temperature), d_{10} = grain size [mm] of the sediment at the intersection with the 10% line of the sum curve. The formulation is only valid if d_{60}/d_{10} is smaller than five (d_{60} is the grain size [mm] of a sediment at the intersection with the 60% line of the sum curve).

Sequential extraction procedure of metals according to (Zeien and Brümmer 1989) was carried out to ascertain that the filter materials do not contain heavy metals in the mobile and easily supplyable phases at critical concentrations. In order to distinguish between the fractions that were (1) mobile, (2) easily supplyable, (3) bound to Mn oxides, (4) organically bound, (5) bound to poorly crystallized Fe-oxides and (6) bound to crystallized Fe-oxides, different reagents were used to extract the filter media (Table 4.3). The residual fraction that is usually included in the Tesier sequential extraction procedure (Tesier et al. 1979) was excluded while it was not the objective of this work to characterize the residual fractions of heavy metals in the filter materials. Samples of 2 g each of filter material were investigated in this way.

Table 4.3. Fractions and extraction solutions used during sequential extraction (Zeien and Brümmer 1989) and (Krüger and Schneider 1994).

Fraction	Extraction solution	Ecological meaning
Mobil fraction	Ammonium nitrate	Mobilization is given by acid input
Exchangeable fraction	Ammonium Acetate	Input of acids and bases: soil-Carbonate buffer (Ulrich 1991)
Mn-oxides fixed fraction	Hydroxil Ammonium Chloride	Soluble during more acid input
Organically bound	Ammonium EDTA	Soluble with solution or precipitation of humic substances
Poorly crystallized Fe bound	Buffer of Ammonium Oxalate	Poor soluble fraction, Al-Fe- soil buffer (Ulrich 1991)
Crystallized Fe-bound	Solution of Ascorbic acid+ Ammonium-Oxalat Buffer	------
Residual Fraction	HNO_3 + $HClO_4$	(Insoluble fraction)

4.3 Column Experiment

4.3.1 Column Set up and Procedures

The column experiment was performed at the Institute of Mineralogy and Geochemistry (IMG), University of Karlsruhe using highly contaminated simulated water that was representative of the pollution scenario in the study area. Thus, Synthetic Polluted Water (SPW) with roughly the same composition as that of the Akaki Rivers and Aba Samuel Lake was prepared in laboratory according to OECD Guidelines for Testing of Chemicals (OECD 1999). Accordingly the following chemicals were dissolved in a liter of tap water in order to prepare the SPW: Peptone (160 mg), Meat extract (110 mg), Urea (30 mg), K_2HPO_4 (28 mg), $CaCl_2.2H_2O$ (4 mg), NaCl (7 mg), and $MgSO_4.7H_2O$. The resulting DOC (2.7 mg/l) value of the simulated water is typical of tropical rivers (Meybeck 1982). Each 30 ml of the SPW solution was added in to a 50 ml PE bottle and kept under freezing in order to avoid biodegradation. In a 5L Erlenmeyer flask initial metal solutions were prepared using 2 PE bottles of SPW (each 30 ml) as background solution mixed with tap water of IMG, University of Karlsruhe (pH 7). No pH adjustment was necessary while the simulated water has similar pH as that of the Akaki waters. In order to simulate the high content of suspended particulate matter (SPM) in the Akaki Rivers and the lake it was necessary to spike the simulated polluted water (SPW) with Na-rich Montmorillonite of Wyoming type. The Akaki River and Aba Samuel Lake are loaded with high-suspended particulate matters (SPM) where montmorillonite is the dominant clay (Alemayehu 2004). The concentration of SPM in Akaki Rivers ranges up to 5 g/l while in groundwater it reaches up to 10 mg/l. Heavy metal concentrations spiked into the SPW were 10^3 µg/l of Cu, Zn, Ni, As and 250 µg/l of Pb prepared from metal standard stock solutions. All working solutions were prepared by diluting the stock solution with double distilled water. Characteristic initial dissolved metal concentrations and some parameters of the simulated polluted water are given in table 4.4. During the column experiment, the sorption of heavy metals by montmorillonite before passing through the columns was also taken into consideration. The net dissolved metal concentrations were used as initial metal concentrations in determining removal efficiency of various media.

Laboratory-scale filter columns were used to test if the natural materials would remove the heavy metals considerably. The simulated polluted water was treated using eight parallel lab-scale down flow filter columns filled with homogeneous materials. The materials were mixed with quartz sand in a ratio 1:10 in order to enhance their permeability during the column experiment. The depth of each column is 30 cm and the internal diameter is 3 cm (Fig. 4.1). A peristaltic pump was used to pump the simulated water through the columns from a 5 liter Erlenmeyer flask at flow rate of 0.12 ml/min in accordance with the DIN V 19736 recommendations for column experiment specifications (Odensaß et al. 2000). A mesh (Ø 0.5 mm) was placed at the bottom of the columns to prevent loss of filter materials. Throughout the column experiment, in order to minimize the early coagulation of the columns by SPM a magnetic stirrer was placed in Erlenmeyer flask. The flow rate, pH, DOC, spiked metal and SPM concentrations were kept constant throughout the experiment period.

Table 4.4 Initial dissolved metal concentration and water parameters of the SPW in µg/l.

Metal	Concentrations (µg/l)
Mn	11.6
Fe	36.1
Ni	640
Cu	490
Zn	737
As	920
Cd	1.75
Pb	76.3
pH	7.0
DOC (mg/l)	0.03
SPM (mg/l)	1000

The filter media used to pack the columns were DQ Scoria, PYC Pyroclastic, KVA volcanic ash, EBF laterite, MnR sediments and QTZ sand.

In the first week of the experiment pH, dissolved oxygen concentrations, and conductivity were measured every 24 hours and effluents were sampled simultaneously. Depending on the results observed in the first week of the experiment, effluents were sampled two times in a week afterwards. Water quality parameters such as dissolved oxygen, electrical conductivity and pH were measured immediately after sampling of effluents using the WTW Multi-Line sensors.

After water parameter measurements, the effluent samples were further pre-treated and preserved before carrying out analytical measurements. For heavy metal analysis, all effluent samples were filtered through 0.45 µm pore size cellulose nitrate filters and collected in separate pre-cleaned 50 ml polyethylene plastic sampling bottles. In order to preserve the samples at pH 2, concentrated suprapur HNO_3 (0.1 ml) was injected in all water samples. Samples were kept in a cool dry place under

+4 °C before they were measured. Unfiltered samples were also put in to 20 ml polyethylene plastic sampling bottles for anionic measurements.

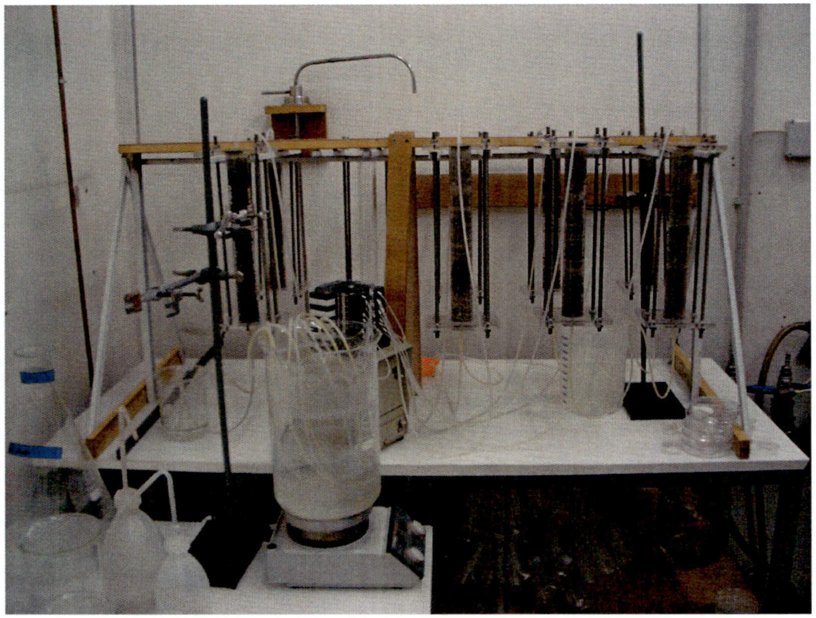

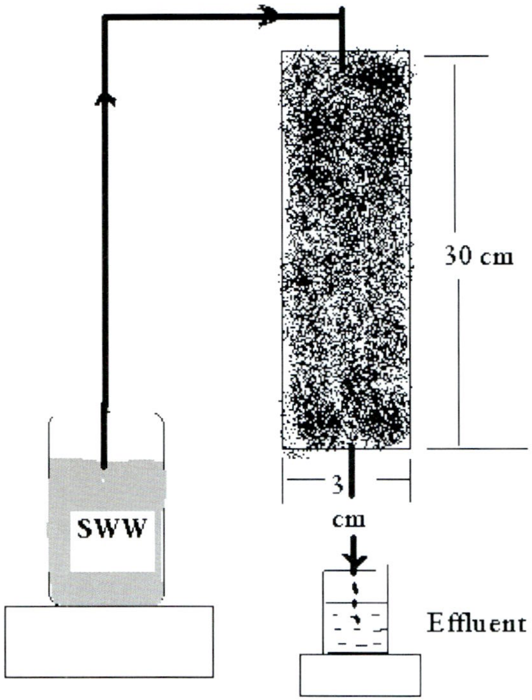

Column Specifications

D = 15 mm; L = 30 mm

Flow Rate = 0.12 ml/min;

Flow: down- flow mode

Fig. 4.1 Column experiment setup.

4.3.2 Sequential Extraction Procedures

Based on the color of the columns sludge, content of organic matter, formation of H_2S (evidenced by foul odor), redox conditions and expected variability in the distribution of heavy metals within the columns, the sludge has been sampled in 3 - 4 depth sections (Table 4.5). Distribution of the montmorillonite spiked within the columns was also one criterion that was considered in depth profiling. The samples were well homogenized before they were used in the leaching and sequential extraction procedures.

Table 4.5. Depth sections of samples of filter media after column experiment (cm).

Column sludge	I	II	III	IV
KVA Volcanic ash	0 - 5	5 - 15	15 - 20	20 - 30
PYC pyroclastic	0 - 10	10 - 20	20 - 30	
MnR sediment	0 - 5	5 - 15	15 - 20	20 - 30
DQ scoria	0 - 5	5 - 15	15 - 25	25 - 30
EBF laterite	0 - 15	15 - 25	15 - 30	
QTZ sand	0 - 5	5 - 15	15 - 30	

Metals speciation in the media spent has been characterized after applying BCR (Community Bureau of Reference) sequential extraction procedure. One gram of dried sample from each filter material was added to a 50 ml centrifuge tube. The samples were extracted using the following extractant solutions according to the method that is well described in (Davidson et. al 1999, 2000).

Step 1. Easily exchangeable, water and acid soluble metals - Acetic acid (0.11 mol/l).
Step 2. Reducible species e.g. bound to Fe/Mn oxyhydroxides - Hydroxylammonium hydrochloride (0.1 mol/l), pH 2.
Step 3. Oxidizable species e.g. bound to organic matter or sulfides - Hydrogen peroxide (8.8 mol/l) and ammonium acetate (1 mol/l), pH 2.

Performance of the BCR extraction procedure

*Step 1.*One gram of each filter media sample spent was extracted with 40 ml of 0.11 mol/l acetic acid per grams of (dry) soil, by shaking for 24 hours at room temperature. The extract was separated from the residue by centrifugation, filtered, decanted, acidified with nitric acid and stored in a polyethylene container at 4 °C. The residue was washed with 20 ml distilled water by shaking and centrifuging. The supernatant was discarded.

Step 2. The residue was extracted for 24 hours with 40 ml/g, freshly prepared 0.1 mol /l hydroxylammonium chloride acidified to pH 2 with nitric acid. The extract was separated and the residue washed as described above.

Step 3. The residue was digested with 10 ml/g, 8.8 mol/l hydrogen peroxide, firstly for 1 h at room temperature and then for 1 h at 85 °C in a water bath. The beaker was covered with a watch glass during the digestion, and was occasionally shaken. At the end of the second hour, the watch glass was removed and the solution evaporated to a few ml. This procedure was repeated with a second volume of hydrogen peroxide. After cooling, the moist residue was extracted for 24 hours with 50 ml/g, 1.0 mol/l ammonium acetate adjusted to pH 2 with nitric acid. The extract was isolated as described before. Trace metal analysis was carried out using HR ICP-MS.

In order to understand how the media spent will react in natural geochemical conditions after being disposed, a leaching experiment using tap water was performed. A single step water extraction was employed to characterize the mobility of water-soluble fractions of metals in the used media. 5 g of media spent was extracted with 50 ml of tap water at pH=7, which represents the natural freshwater geochemical conditions for 24 hours. The extract was separated from the residue by centrifugation, filtered through folded filter paper, decanted, acidified with nitric acid and stored in a polyethylene container at 4 °C. Trace metal analysis was carried out using HR ICP-MS.

4.4 Analytical Technique and Data quality

4.4.1 Water Sample Analytical Procedures

4.4.1.1 High Resolution Inductively Coupled Plasma-Mass Spectrometry (HR-ICP-MS)

Heavy metal concentrations of Sc, V, Cr, Mn, Fe, Co, Ni, Cu, Zn, As, Rb, Sr, Mo, Ag, Cd, Sn, Sb, Ba, Tl, Pb, Bi, Th, and U were measured using High Resolution Inductively Coupled Plasma Mass Spectrometer (HR-ICP-MS) (Axiom, VG) using external calibration with certified multi-element standards. Accuracy was checked in each analytical batch by measuring the certified reference material 'Trace Metal in Drinking Water' (TMDW) from High Purity Standards (HPS, USA), and was found to be within ± 5% of the certified values. To correct for instrumental drifts and non-spectral affects Rh, In and Th were used as internal standards. Precision was evaluated by the standard deviation of three individual runs, and was better than 2% (1s). The specific working conditions of the HR ICP-MS are shown in table 4.6.

4.4.1.2 Total Reflection X-ray Fluoresence Analysis (TRFA)

Suspended particulate matter (SPM) samples collected on 0.45 µm pore size cellulose nitrate filters (Sartorius AG, Germany) were further treated at IMG laboratory for heay metals concentration. Each filter was digested with concentrated HNO_3 on a heating plate at an appoximate temperature of 250 - 300 °C. The solution was let to evaporate slowly until the cellulose filter was completely dissolved and a relatively homogeneous solution was attained. Analysis of heavy metals in the prepared solution was carried out using Total Reflection X-ray Fluoresence Analysis (TRFA).

A drop of the prepared solution (20 µl) was placed on a silica carrier. On evaporation of the solvent a thin film, a few nanometers thick, remains. In practice, the greater part of the scattering normally arising from the sample and its matrix is eliminated. The simplified quantitative analysis has been made possible after adding a 5 µg Rubidium internal standard into each sample solution considered. By illuminating the sample with a beam that is being totally reflected, absorption of the beam in the supporting substrate is largely avoided and the associated scattering was greatly reduced. This also reduced the background noise substantially. A further contribution to the reduction of the background noise is obtained by minimizing the thickness of the sample.

Table 4.6. Working parameters of the HR ICP-MS.

Instrument	VG Elemental (Axiom)	
Forward power	1350 W	
Reflected power	< 20 W	
Sampling uptake rate	1 ml/min	
Scan mode	Magnet scan	
Torch	Quartz torch (Glass Expansion, Australia)	
Nebulizer	Type Meinhardt (Glass Expansion, Australia)	
Spray chamber	Conical impact wall (bead)	
Temperature of spray chamber	10±1	
Argon gas flow rates	Cool gas:	14.2
	Auxiliary gas:	0.65
	Nebulizer gas:	0.94
Data acquisition parameters		
Acquisition mode	Pulse counting	
Number of runs	3	
Dwell time	25 ms	
Points per peak width	20	
Sweep	1	
Internal standard	103Rh	

Determination of concentrations of heavy metal in samples Ku1-Be8 was made by standardizing the Fe concentration that was already measured by FAAS. This is because of the relatively high concentration of Fe in these SPM samples. Since the relatively low Fe concentrations in samples Sd9-Ab22 could not be used as standard, Rb standard was used to determine concentrations of heavy metal of interest in these SPM samples. Results of Fe, Mn, Cu, Pb measurements indicate good precision. The relative standard error in quantifying Fe, Cu, and Zn bound to SPM in these samples were around 13% while Ni and As showed varying error values (up to 20%) that might have been caused by errors in sample preparation. The detection limit was calculated for a representative sample and found to be better than 4%.

4.4.1.3 *Flame Atomic Absorption Spectrometry (FAAS)*

At IMG laboratory, the concentrations of Na, K, Mg, and Ca were measured in the water samples with a Perkins - Elmer Flame Atomic Absorption Spectrometry (FAAS) 1100B. Before measuring the samples, a 5% La_3O_3 solution was prepared along with standard solutions of metal ions of interest. A five-point calibration was used all through the measurements. The metal standards were prepared from 1000 mg/l standard stock solutions as in concentration ranges shown in table 4.7. A blank solution was measured along side with each metal cation under consideration. Table 4.7 presents the ranges of standard calibration solutions used. The concentrations of Mg and Ca were determined after DIN EN ISO 7980 (2000) while DIN 38406 - 13 and 14 (1992) was followed to determine those of K and Na.

Table 4.7. Metals stock and standard solutions used.

Metal	Metal stock solution (mg/l)	Standard calibration solutions (mg/l)
Na	1000	0.05; 0.1; 0.25; 0.5
K	1000	0.05; 0.1; 0.25; 0.5
Mg	1000	0.05; 0.1; 0.25; 0.5
Ca	1000	0.5; 1.0; 2.5; 5.0

The possible matrix interference effect from aluminum was first checked with a help of mixture of NO_2 gas and acetylene before measuring the metal ions in the water samples. Concentrations determinations were made after drawing calibration curves that were prepared from the respective metals standard solutions. The relative standard deviation in all measurements were below 5%

4.4.1.4 *Ion Chromatography (IC)*

Concentrations of nitrate, sulfate and chloride in surface water and groundwater samples were measured at IMG Laboratory with Metrohm 690 Ion Chromatograph. The eluent used for separation of chloride, sulphate and nitrate (anions) was prepared by adding 4.15 g of pthalic acid, 10 ml of acetonitrile and 950 ml of distilled water. In order to get the salt ionized completely, the pH was adjusted to 3.8. Then distilled water was added to make the total volume to 1 liter. A sub sample of 100 ml was taken from this solution and 50 ml of acetonitrile was added and again made to 1 liter.

Before injecting the sample into the chromatograph, the eluents were pumped through the ion exchange columns so that the resin exchange sites could contain the eluent ions. The samples were diluted 5 to 10 times, depending upon the concentration of Cl^- to facilitate the measurement of other anions and cations present in lower concentrations.

4.4.1.5 *Scanning Electron Microscope (SEM)*

In order to characterize the morphology and chemical compositions of the suspended particulate matter (SPM), the filtered particles of the water samples have been analyzed by Scanning Electron Microscopy (SEM) LEO 1530 Gemini at the University of Karlsruhe. For this analysis, SPM samples were prepared by wetting them using C instead of Pt and accelerating voltage was 15 kV. The cellulose filter paper was spread on a double-sided adhesive tape held on a sample holder. Later it was coated with carbon to minimize charges buildup and was examined under the SEM. Different photomicrographs were taken at a magnification of 1000X. Relatively magnified photomicrographs were then taken based on features observed at different areas of the matrix.

4.4.2 Filter Media Analytical Procedure

4.4.2.1 *X- Ray Diffraction (XRD)*

Five filter media samples of different sources have been analyzed by XRD. Initially bigger aggregates were crushed and homogenized. The samples were dried at 105

°C in an oven for about 24 hours then were milled with a Labotech milling machine to less than 3 µm. In order to get a preferable crystal orientation during analysis, the specimens were smeared into a plastic sample holder before measurement. The specimens were well homogenized before they were pasted on the plastic sample holder. Analysis of the prepared specimens was carried out using a Simens (D500) diffractometer, which applies a Cu target (wavelength for the K1 radiation: 1.5418 Å). The scanning range is between 3° to 63° at a speed of 0.5°/minute, with a step width of 0.01°. During analysis, the measurement conditions were kept at 45 kV, 25 mA. Mineral identification was done by considering the 'd' spacings, after applying the Bragg's law (nλ = 2d siṇ, where 'd' is the distance between successive atomic plane of a certain crystal, λ is the monochromatic wavelength of the incident beam and θ is the incident angle) (Warren 1981).

4.4.2.2 X - ray Fluorescence Analysis (XRF)

Measurement of heavy elements was carried out using energy dispersive fluorescence (EDXRF) that applies electronic sorting of pulses of different amplitude produced by x-ray photons of different energies (Williams 1987). Whole filter material samples were prepared by using procedures similar to those described for EDXRF analyses of silicate minerals. Samples were ground to fine powders using an agate mortar. The samples were kept at constant particle size to avoid the risk of variation in fluorescence intensities. The powder samples were put in plastic sample cups and were covered by thin plastic sheets very tightly with the help of plastic bands. The covered surface was kept very smooth to avoid scattering of X-rays. All intensity measurements were made on a Spectrace 5000 spectrometer. The samples were measured against the soil standard GXR-2 (Govindaraju 1994) using Cu and Pd filters in Spectrace 5000 spectrometer. Details on the analytical procedure and detection limits for the bulk samples are referred to Kramar (1997).

Determination of the major rock-forming elements in the filter media was carried out using wavelength dispersive x-ray fluorescence (WDXRF), which applies crystals to disperse the component wavelength of the X-ray spectrum emitted from the samples. The WDXRF technique utilizes a lithium tetraborate fusion disc sample for presentation to a 13-channel simultaneous wavelength dispersive x-ray fluorescence spectrometer.

Glass transformed samples were prepared for WD-XRF analysis by mixing 1 part of dry sample with 4 or 10 parts of lithium tetraborate flux. After properly mixing, they were heated using gas burners for a period of 30 minutes. Heating was done in platinum crucibles. During fusion, the melt was agitated for the production of homogeneous glass. The glass-transformed samples are favored for the measurement of major element, as the decomposition of a portion of sample and flux produces a homogeneous glass and it eliminates the particle size and the mineralogical effects (Tertian and Claisse 1982). The ideal specimen shape for XRF measurement is a flat polished disc and this was obtained by casting the molten glass in a circular mould of required size. The samples were measured against the BE-N standard in the SRS 303 AS instrument. The measured standard deviations of the major elements were between 1 - 3 mass % while for trace elements it was generally below 5 ppm.

4.4.2.3 Carbon sulfur analysis (CSA)

Leybold 5003 carbon sulfur analyzer (CSA) was used to measure total carbon and sulfur in percentage. It was calibrated with respect to the 0.42% carbon containing free cutting steels. The other constituents of the standard are sulfur (0.147%), silica (0.46%), phosphorous (0.031%), and manganese (1.15%). For the analysis, approximately 100 mg of dry bulk sample was added to approximately 500 mg of iron chip accelerator. To this mixture, tungsten (wolfram) was added such that the ratio between tungsten accelerator and iron cheap accelerator was 3:1. The instrument oxidizes the carbon and sulfur present in the sample with a constant flow of O_2. The detection limit was estimated to be 0.01%.

4.4.2.4 Carbon water analysis (CWA)

Organic carbon in the bulk samples was estimated through measurement of total carbon and inorganic carbon content by Leybold 5003 carbon water analyzer (CWA). By calculating the difference between values of the total carbon and the inorganic carbon, the content of organic carbon was determined. For inorganic carbon analysis, the instrument was calibrated with respect to $CaCO_3$. Approximately 250 ml of dry sample was used for the measurements. The sample was burnt with a steady flow of N_2 and the carbonates were converted into CO_2. The output was obtained in terms of percentage of CO_2. The relative standard deviation measured was below 2%.

5 Results

5.1 Akaki Water Samples

Water parameter measurements taken on site have shown that most groundwater samples have pH values ranging between 6.5 and 7.5 while values are variable in the surface water samples (6.7 - 8.0). Comparison of water parameters of surface and groundwater samples collected at the two different sampling seasons is presented in appendices I & II. The Akaki Rivers and Gerado pond water have shown relatively higher pH values than the Aba Samuel Lake (Table 5.1). These pH values are in the acceptable range of water quality criteria set by US EPA. Other water parameter measurements also indicate that most groundwater samples show variable dissolved oxygen concentration and conductivity depending on the proximity of these water sources to the polluted surface water. Specific conductance (Table 5.1 and 5.2) indicate low dissolved ion concentrations upstream from the Akaki Rivers, and an increase in the lower lying Aba Samuel Lake area. Most of the surface water samples show lower dissolved oxygen amounts as compared to the groundwater samples. The polluted Akaki River at sample site Ku1 shows the lowest dissolved oxygen concentrations (0.2 mg/l).

5.1.1 Water Types

Results of chemical analysis on the surface water and groundwater samples are presented in table 5.1, 5.2, and appendices I & II. Concentrations of Na, K, Mg, and Ca are compared between selected water sources that are in close connection with the Aba Samuel Lake. The dissolved concentrations of Na, K, Mg, and Ca in Akaki River at sample site Ku1 were 62 mg/l, 26 mg/l, 15 mg/l, and 38 mg/l respectively, which are similar to concentrations in Aba Samuel Lake at Dk4 (Table 5.1). In surface waters, the highest Na and K concentrations are measured in the Akaki River at El18 (ELICO). Concentrations of Na up to 86 mg/l and K 40 mg/l are measured at El18. Concentrations of Mg and Ca are similar in the different samples taken from the Akaki River (Table 5.1).

The Small Akaki River at El18 showed relatively higher concentrations of all major cations except Mg compared to that of samples taken downstream of the river. As compared to the other surface water samples, the Akaki Textile Factory (Ak14) wastewater samples show the lowest K concentrations (Table 5.1).

The major cationic compositions of surface water and groundwater samples with their respective location in the study area are presented (Fig. 5.1a and 5.1b). Upstream of the Small Akaki River (El18) the major cations are relatively higher than the Aba Samuel Lake and the outlet. This can be explained by the high waste effluents discharged by the ELICO Tannery and other industrial sources. Concentrations of Na, K, Mg, and Ca are however similar among the Aba Samuel Lake (Dk4) and the outlet of the lake (Dk5) samples (Fig. 5.1).

Table 5.1. Water parameters, major cationic and anionic compositions of surface water.

Sample	O_2	T	pH	Cond.	Na^+	K^+	Mg^{2+}	Ca^{2+}	Cl^-	NO_3^-	SO_4^{2-}
	mg/l	°C		µS/cm				mg/l			
Ku1*	0.26	21.0	6.9	437	62.0	26.0	15.0	37.8	31.4	1.8	17.9
Dk4	2.86	22.1	7.2	756	50.0	29.0	16.0	44.0	54.6	0.0	40.1
Dk5	2.42	20.1	7.4	611	46.0	15.0	14.0	40.0	45.6	1.0	28.2
Gr7	8.49	29.0	8.0	523	14.0	47.0	23.0	27.5	10.9	0.0	7.4
Ju11*	3.25	19.0	7.3	516	38.0	20.0	40.0	26.0	46.6	14.8	43.7
Ak14	0.4	26.0	7.3	1029	68.0	5.0	19.5	70.5	57.6	0.0	71.9
EL18*	0.04	21.0	7.5	1058	86.0	40.0	20.0	56.0	46.6	14.8	43.7
EL19	0.06	27.0	6.8	1037	29.0	4.0	24.5	32.0	46.6	14.8	43.7

**Samples were taken from downstream - upstream direction in the Small Akaki River.*

On the other hand, the concentrations of Na and Ca in most groundwater samples are higher as compared to K and Mg (Table 5.2). In groundwater samples, higher Na and K concentrations are measured at Dewera Guda (Ge2) water well. Except for De2 (18 mg/l) and Tl12 (10 mg/l) there are no significant differences in concentrations of K between the groundwater samples. In almost all groundwater samples, K concentrations are below 10 mg/l. The highest Na and K concentrations in groundwater samples are measured at De2 (86 mg/l and 18 mg/l respectively). The lowest concentrations are measured at Dk3 (3 mg/l) (Table 5.2). High concentration of Na up to 84 mg/l is measured in groundwater at Ls16. Concentrations of Mg and Ca are variable in most groundwater samples (Table 5.2). Calcium concentrations are the highest in groundwater samples taken from IV13 and Ls16 (74 mg/l and 42 mg/l respectively). In most of the groundwater samples, the concentrations of Mg are relatively higher than that of the surface water samples except at sample site Ju11.

Table 5.2. Composition of major cations in the groundwater samples including samples from the Akaki well fields.

Sample	O_2 mg/l	T °C	pH	Cond. µS/cm	Na^+	K^+	Mg^{2+}	Ca^{2+}	Cl^-	NO_3^-	SO_4^{2-}
					mg/l						
De2	2.13	29	7.28	676	86.00	18.00	13.00	39.50	36.87	3.78	91.13
Dk3	3.63	25	6.70	410	6.00	3.00	14.00	43.50	39.48	2.79	91.11
Be6	3.54	30	7.33	678	82.00	9.00	13.00	47.75	-	-	-
Be7	5.64	22	7.30	773	19.00	3.00	22.50	38.50	5.43	13.46	6.20
Sd9	5.55	21	7.58	625	26.00	4.00	25.00	49.50	15.76	43.26	17.31
E10	4.97	30	6.90	585	27.00	3.00	24.00	49.00	7.02	21.18	10.96
Tl12	2.17	22	6.98	563	66.00	10.00	39.00	55.00	4.80	5.67	9.42
Iv13	0.77	25	6.60	852	28.00	6.00	42.00	74.00	34.33	18.68	36.98
Ah15	1.37	22	7.01	866	28.00	5.00	21.00	73.00	36.37	19.17	36.02
Ls16	3.82	25	7.35	609	84.00	7.00	22.50	42.00	5.04	8.58	10.14
Ak17	5.4	20	7.56	560	50.00	6.00	23.00	30.00	8.41	11.34	12.20
Td20	6.34	23	7.16	528	20.00	3.00	32.00	32.50	6.82	20.79	10.30
Am21	5.65	23	7.15	475	25.00	4.00	16.00	43.50	5.45	13.23	63.30
Ab22	5.26	23	7.29	512	11.00	3.00	8.50	21.25	5.37	16.38	8.05

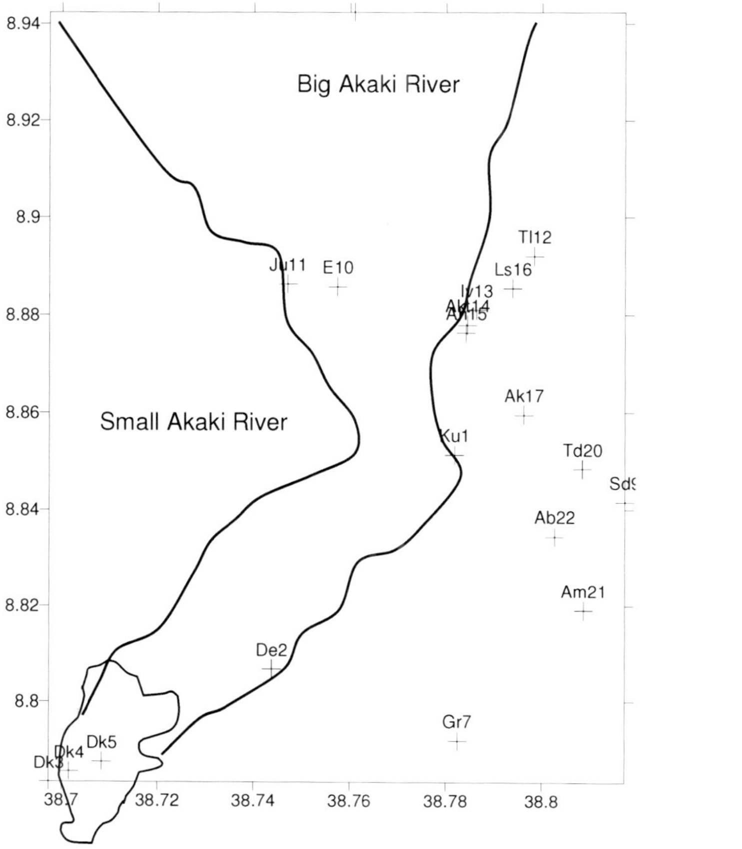

Fig. 5.1a. Schematic of the Akaki Sub-Basin.

b. Cationic compositions of surface water samples.

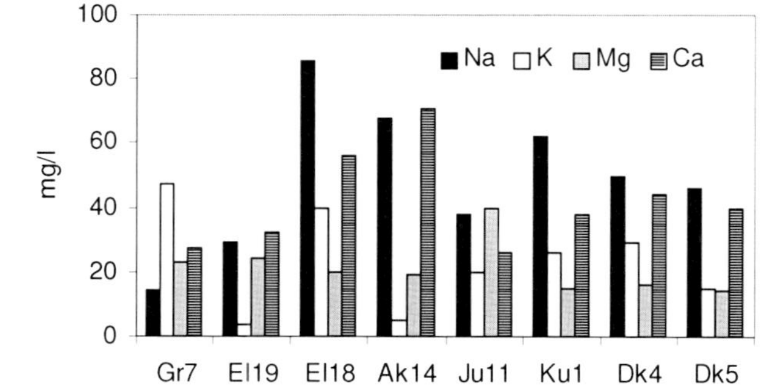

c. Cationic compositions of groundwater samples.

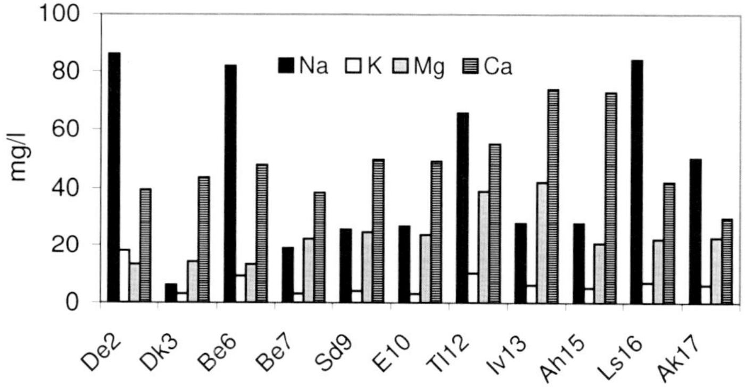

The Little Akaki River at Ju11 and El18 show much higher concentrations of Cl^-, NO_3^- and SO_4^{2-} than those of the groundwater but are lower as compared to the effluents of the Akaki Textile Factory at Ak14 (Fig. 5.2).

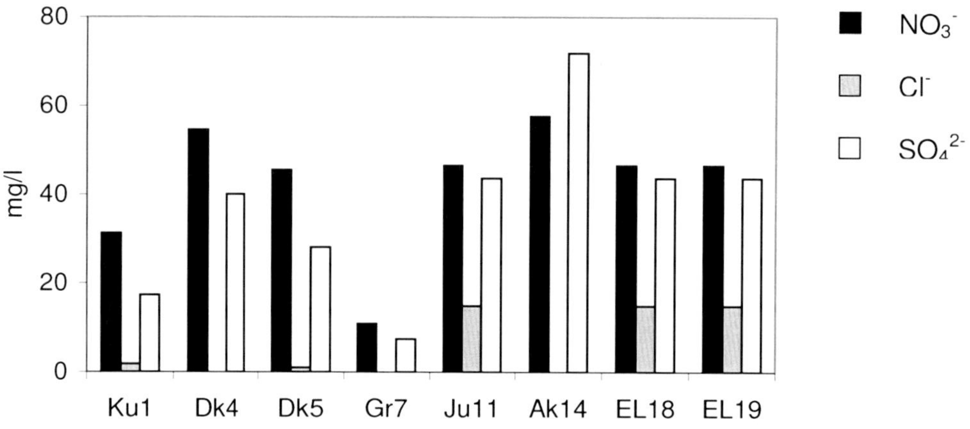

Fig. 5.2. Concentrations of Cl^-, NO_3^- and SO_4^{2-} during the 2003 field campaign were highest in surface water as compared to groundwater samples.

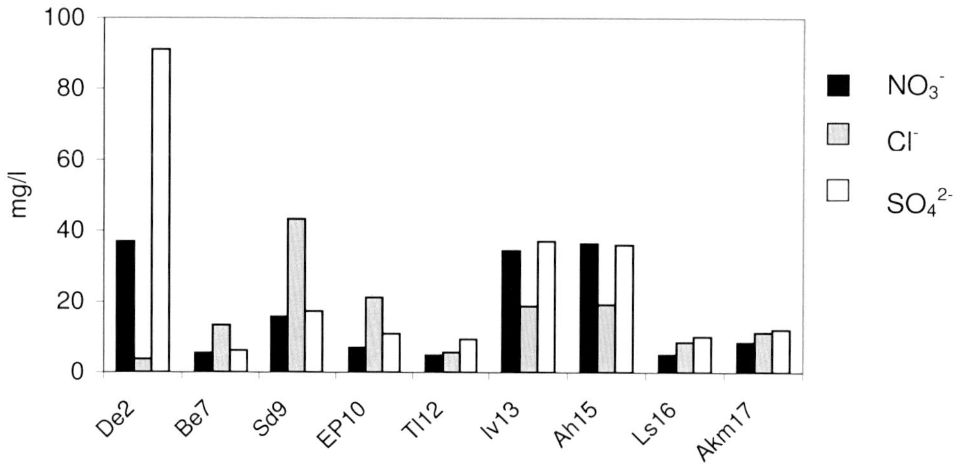

Fig. 5.3. Concentrations of Cl^-, NO_3^- and SO_4^{2-} in groundwater samples during the 2003 field campaign.

Only NO_3^- was found in concentrations exceeding the U.S. Environmental Protection Agency (US EPA) maximum contaminant level (MCL). The concentrations of NO_3^- in groundwater samples are higher than the USEPA MCL (10 mg/l) in 50% of all water samples collected (Fig. 5.3). On the other hand, the concentrations of NO_3^- in surface water samples are two times higher than the USEPA MCL (10 mg/l) in 87% of the samples collected. Industrial effluent sample collected from the Akaki Textile Factory wastewater at Ak14 show high Cl^- and SO_4^{2-} concentrations (60 mg/l and 72 mg/l respectively) while NO_3^- concentration is close to 0 mg/l (Fig. 5.2).

Results of hydrochemical analysis of the surface water and groundwater samples show that the type of water in the study area is alkaline earth metal fresh water with higher contents of alkali ions and prevailing hydrogen carbonate.

From the piper diagram (Fig. 5.4) it can be observed that the surface water and groundwater plot in the same region. However, the surface water samples are evenly dominated by the alkali and alkaline earth metals and hydrogen carbonate while they are slightly enriched in Cl^-, NO_3^-, and SO_4^{2-} anions. On the other hand, the groundwater samples are mainly composed of alkaline earth metals and prevailing HCO_3^- anion. The groundwater samples are also slightly enriched in Cl^- and SO_4^{2-} anions. Regarding the concentrations of anions, HCO_3^- is the dominant one in both surface and groundwater samples.

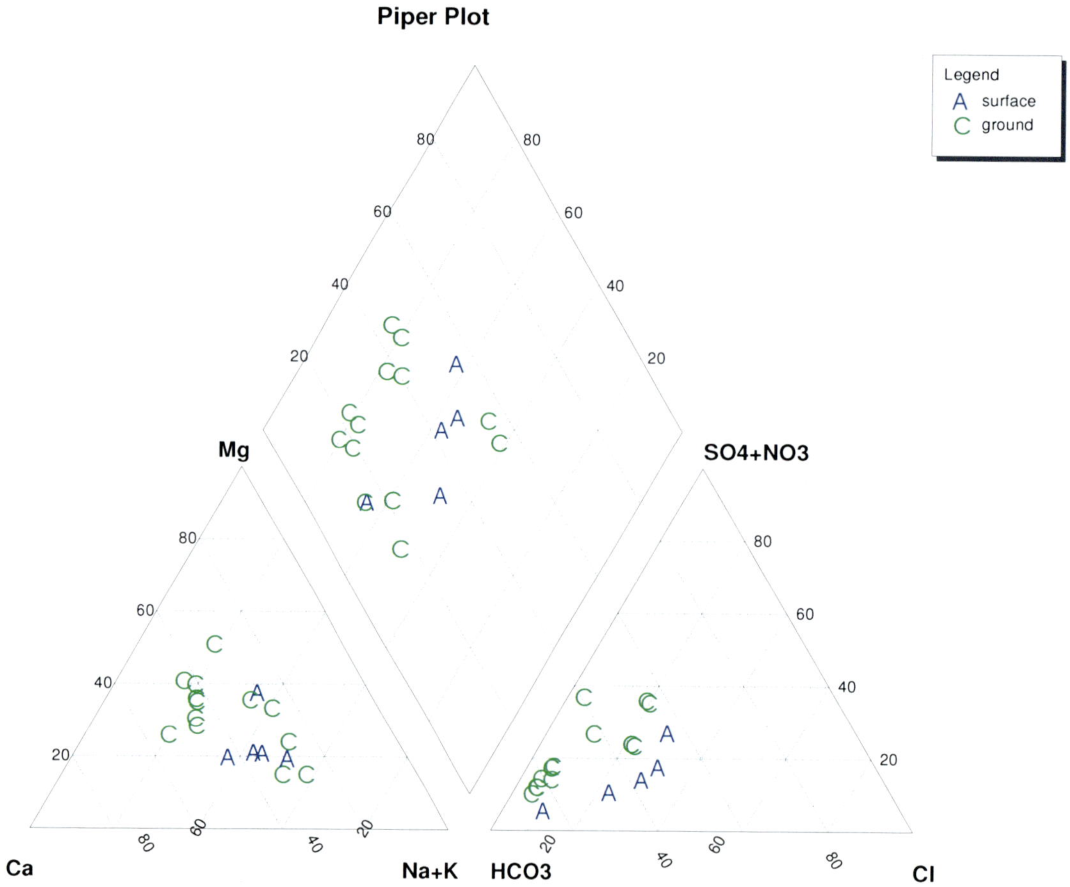

Fig. 5.4. Water type classification in a Piper Diagram.

5.1.2 Dissolved Heavy Metals and Arsenic

In water samples collected in 2003 and 2005, dissolved cations and several heavy metals are detected more commonly and in higher concentrations at sites in areas around the Akaki Rivers and Aba Samuel Lake than at sites in the Akaki well fields and upstream part of the sub-basin. Iron, Mn, Co, Ni, Cu, Zn, Pb and U concentrations are higher in samples from sites downstream of Addis Ababa compared to those of the upstream. Arsenic concentrations are not higher both in surface water and in groundwater samples. Concentrations of heavy metals and As generally increase in water samples that are taken close to the Akaki Rivers and the Aba Samuel Lake.

Concentrations of dissolved heavy metals are close to and above the US EPA MCL (US Environmental Protection Agency Maximum Contaminant Level[1]). In the Akaki River at Ku1 the dissolved concentrations of Cr, Mn, Fe, Co, Ni, Cu, Zn, As, Cd, Ba, and Pb in µg/l are 0.36, 352, 1.66, 5.46, 2.89, 174, 1.40, 0.03, 95.7, and 0.19 respectively (Fig. 5.5 and 5.6). Comparisons of heavy metal concentrations between samples collected during the two sampling campaigns are presented in appendices I&II.

Relatively high dissolved Zn concentrations are measured in surface water samples at Gr7, E10, Tl12 and Ls16. In these samples, dissolved concentrations of Zn reach up to 1.3, 1.6, 2.7, and 3.2 mg/l respectively. Despite the fact that the concentrations were very close to the secondary standards[2] set by US EPA, the continuing improper industrial waste release in to the open water system can increase the concentrations to critical values.

The dissolved concentrations of Fe and Mn in groundwater samples are also well above the secondary standards that are set by the US EPA - 0.3 and 0.05 mg/l for Fe and Mn respectively (Fig. 5.6). In the Aba Samuel Lake, dissolved Mn concentrations are five fold higher than the secondary standards set by US EPA. These concentrations are even ten fold higher than the limiting standards at sample site Gr7. Dissolved concentrations of Fe and Mn are also higher in the Small Akaki River at Ju11, about 90 µg/l and 834 µg/l respectively.

Comparison of the dissolved concentrations of heavy metals and As between the Aba Samuel Lake (Dk4) and its outlet (Dk5) show that concentrations are always higher in the lake than in the outlet. In the lake, the dissolved concentrations of Mn, Fe, Ni, Cu, Zn, As, Cd, and Pb are 2835 µg/l, 98.3 µg/l, 9.7 µg/l, 2.1 µg/l, 43.4 µg/l, 2.8 µg/l, 0.02 µg/l, and 0.14 µg/l respectively. Since the residence time in the lake is long enough for most transport controlling reactions such as oxidation-reduction to take place and the fact that the lake has very low dissolved oxygen, metals can be

[1] Maximum Contaminant Level (MCL) - The highest level of contaminants that is allowed in drinking water.

[2] National Secondary Drinking Water Regulations or secondary standards are non -enforceable guidelines regulating contaminants that may cause cosmetic effects (such as skin or tooth discoloration) or aesthetic effects (such as taste, odor, and color) in drinking water.

released due to anaerobically enhanced reactions. Complex formation with organic ligands may also be responsible for the relatively higher dissolved heavy metal and As concentrations.

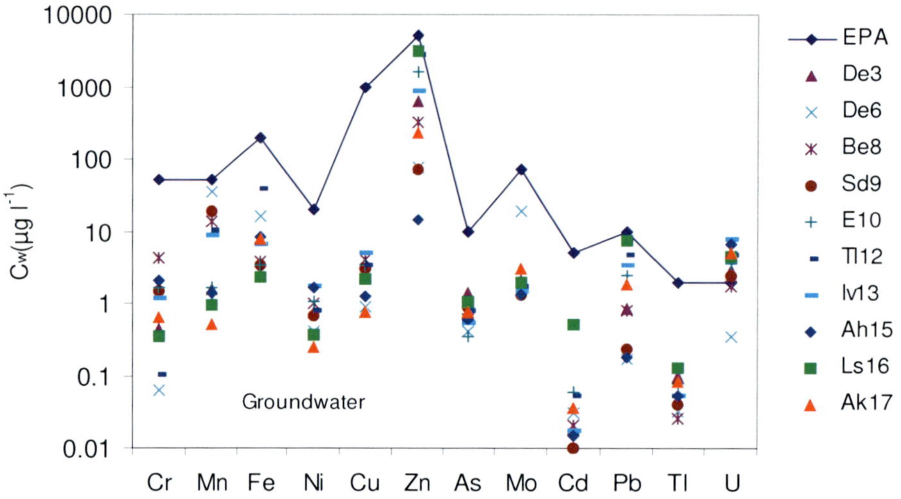

Fig. 5.5. Dissolved metal concentrations in relation to US EPA MCL standards for heavy metals and arsenic (groundwater).

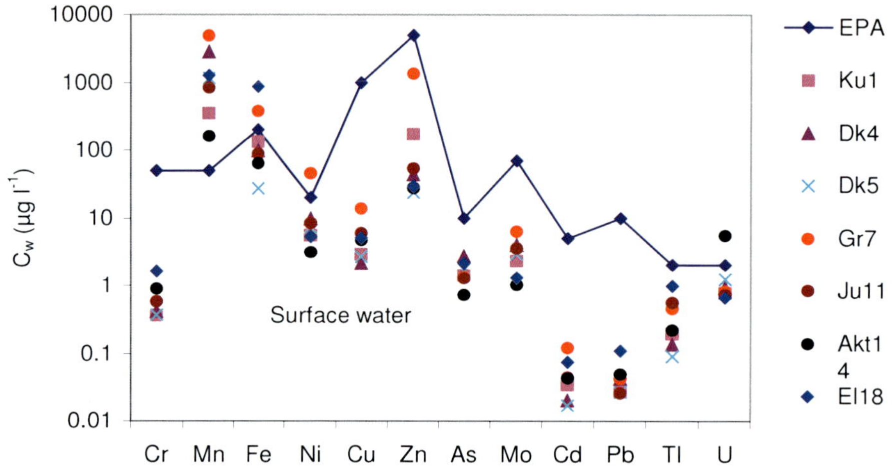

Fig. 5.6. Dissolved metal concentrations in relation to US EPA MCL standards for heavy metals and arsenic (surface water).

In the Aba Samuel Lake outlet (Dk5) the dissolved concentrations of Mn, Fe, Ni, Cu, Zn, As, Cd, and Pb reach 0.37 µg/l, 1153 µg/l, 27 µg/l, 2.33 µg/l, 7.32 µg/l, 2.7 µg/l, 23.9 µg/l, 1.73 µg/l, 0.02 µg/l, 120 µg/l, and 0.09 µg/l respectively. This could be explained by the fact that the lake outlet acts as an oxidation chamber for most of the heavy metals and As that lead to low dissolved concentrations.

5.1.3 Heavy metals Bound to Suspended Particulate Matter (SPM)

Heavy metals bound to SPM in surface and groundwater vary in wide ranges. Concentrations up to 37,000 µg/g Fe, 3000 µg/g Cu, 18,000 µg/g Zn, 2000 µg/g As, and 2000 µg/g Pb are measured in the suspended particulate matter (SPM) of groundwater sample at Dk3 (Table 5.3b). These represent the colloidal fractions of heavy metals in the drinking water wells.

Concentrations of Fe and Mn bound to SPM in groundwater samples are higher at sample site E10 (140,000 µg/g and 6000 µg/g respectively). Iron as high as 150,000 µg/g and 400,000 µg/g are also measured from the suspended particulate matter at sample sites Ah15 and Ak17 (Table 5.3b). This is an indication that in the Akaki Sub-Basin most heavy metals are associated with the suspended particulate matter.

High As concentrations up to 500 µg/g, 200 µg/g, and 120 µg/g are measured in SPM of groundwater samples at site E10, Ah15, and Ab22 respectively. At the same sites Pb 500 µg/g, 197 µg/g and 121 µg/g are also measured (Table 5.3b). At Dewera Guda (De2) and Dika village water well (Dk3) concentrations of As bound to SPM reach up to 883 µg/g and 1413 µg/g. These values rise up to 2400 µg/g in the SPM of Gerado pool (Gr7). Compared to the groundwater samples, the Akaki River and Aba Samuel Lake show relatively lower concentrations of As bound to SPM, the lowest being at sample sites Dk5, Ju11, Akt14, and El18 (Table 5.3a). However, relatively elevated concentrations of As in SPM of surface water samples are found only in samples taken from the Aba Samuel Lake at Dk4 (307 µg/g).

Table 5.3a. Concentrations of heavy metals bound to SPM in surface water (µg/g).

Sample	Fe	Cu	Zn	Mn	As	Pb	Ni
Ku1	74213	39.8	626	1200	10.3	nd	11.7
Dk4	314000	434	3853	12020	307	267	389
Dk5	16000	18.5	249	837	21.3	19.8	16.00
Gr7	6187	5.34	198	221	1.36	nd	8.16
Ju11	8361	4.46	70.0	137	1.45	nd	0.98
Ak14	1433	2.53	17.4	53.7	0.84	1.17	1.71
El18	21.8	5.74	6.84	0.91	1.22	nd	0.89

Table 5.3b. Concentrations of heavy metals bound to SPM in groundwater (µg/g).

Sample	Fe	Cu	Zn	Mn	As	Pb	Ni
Dk3	36750	2613	15403	920	838	965	1198
De6	54267	5303	13775	8034	910	1299	1012
Be8	16364	4212	5011	593	558	679	712
Sd9	1623	68.33	441	69.3	125	73.0	51.2
E10	138175	5846	37574	5902	10611	6216	4363
Tl12	2515	18.23	518	165	40.6	40.63	4306
Iv13	1557	111	646	41.8	30.2	42.86	44.29
Ah15	151373	187	2274	7810	195	196.94	218
Ls16	1787	172	447	257	51.0	46.0	35.5
Ak17	394017	347	12594	13991	68	0.001	477
Td20	1391	67.1	379	70.6	115	67.9	60.7
Am21	19800	57.1	1323	101	48.6	69.3	54.3
Ab22	2968	155	398	118	118	122	138

nd = not detected

Both in surface water and groundwater samples Ni concentrations bound to SPM are variable. In surface water the maximum Ni bound to SPM is measured in Aba Samuel Lake at Dk4 while the lowest is in the Small Akaki River at El18 (Table 5.3a). Compared to the surface water samples, groundwater samples at E10 and Tl12 show the highest Ni concentrations bound to SPM (4360 µg/g and 4300 µg/g respectively). In groundwater, the lowest Ni concentration bound to SPM is measured at sample site Sd9 (50 µg/g).

The concentrations of Cu, Zn, and Pb bound to SPM are compared to the (Bund/Länder-Arbeitsgemeinschaft Wasser (LAWA) 2000) classification. Table 5.3 shows the LAWA classifying metal contamination of SPM in water samples (Irmer

1997). The comparisons are presented in figure 5.7 and 5.8 for Cu, Zn, and Pb in surface water and groundwater samples. For As and Ni there are no LAWA classes available. In comparison with the LAWA classes, the concentrations of Cu, Zn, and Pb bound to SPM are well above the LAWA class I that shows no contamination.

Table 5.4. The LAWA classification of heavy metals bound to SPM (Irmer, 1997).

LAWA	Classes	Cu	Zn	Pb
IV	Very strong contaminated	640	3200	800
III-IV	Strongly to very strongly contaminated	320-640	1600-3200	400-800
III	Strongly contaminated	160-320	800-1600	200-400
II-III	Moderately contaminated to strongly contaminated	80-160	400-800	100-200
II	Moderately contaminated	40-80	200-400	50-100
I-II	Unpolluted to moderately contaminated	20-40	100-200	25-50
I	Uncontaminated	<20	<100	<25

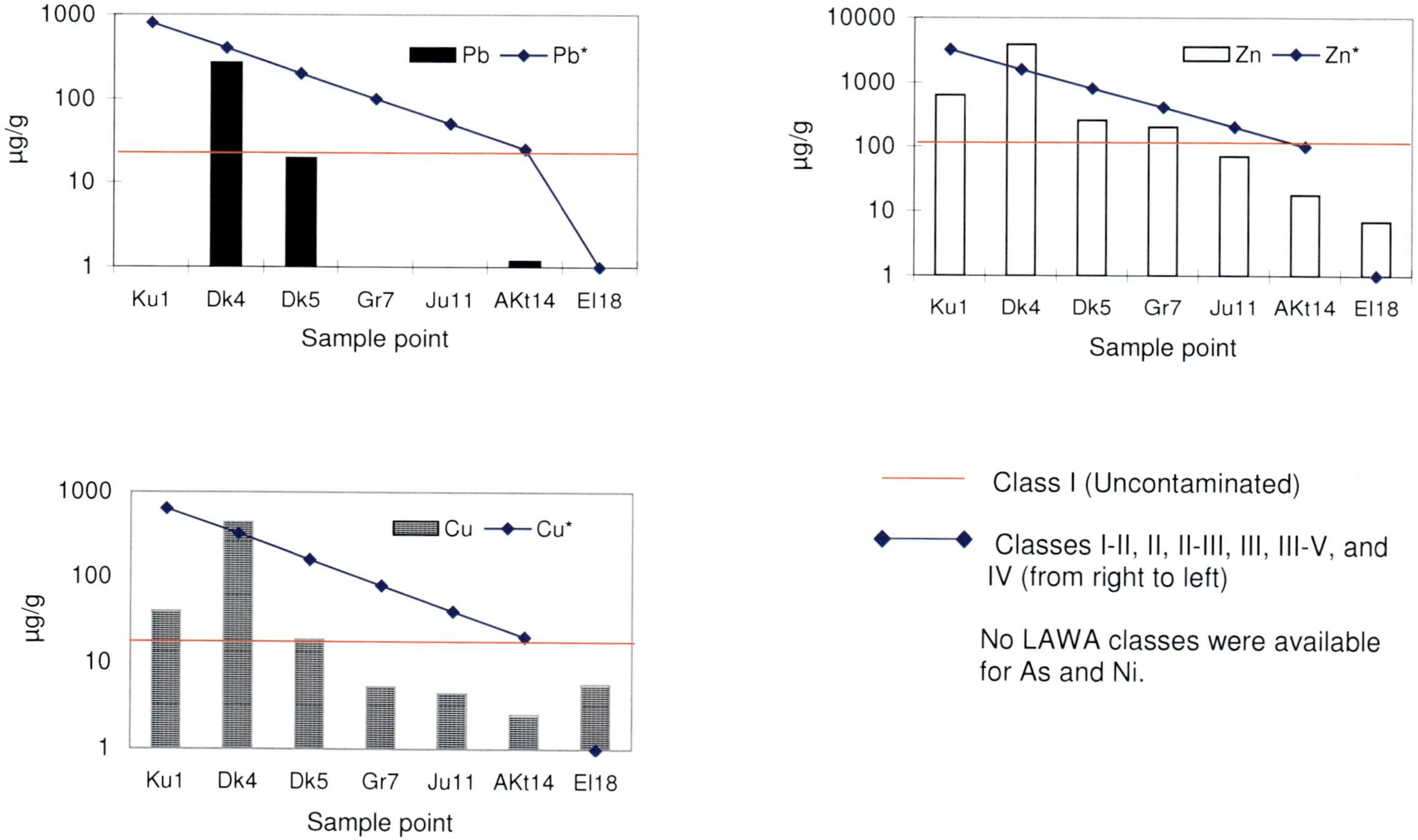

Fig. 5.7. Comparison of concentrations of Cu, Pb, and Zn bound to SPM based on the LAWA classes (*), from the "Very strongly contaminated" left upper corner to the "Uncontaminated" lower right corner (surface water).

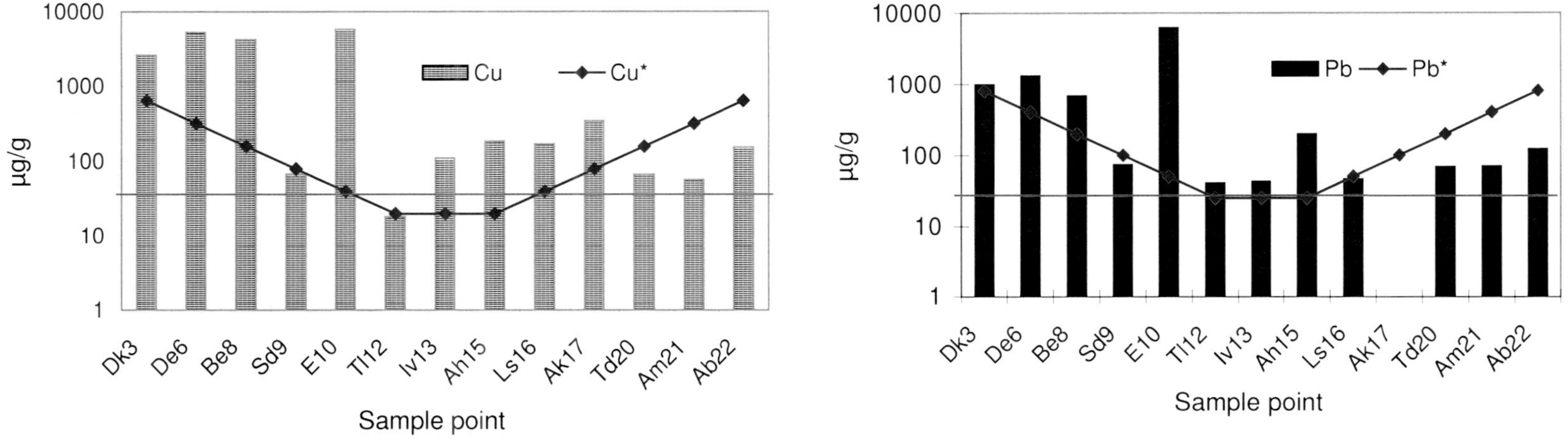

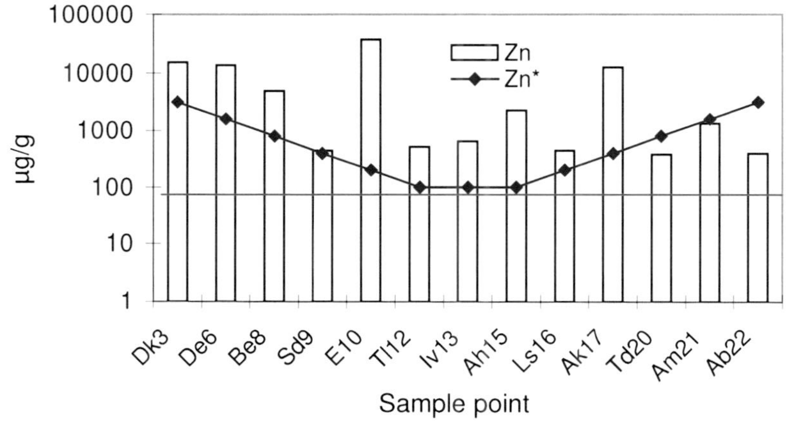

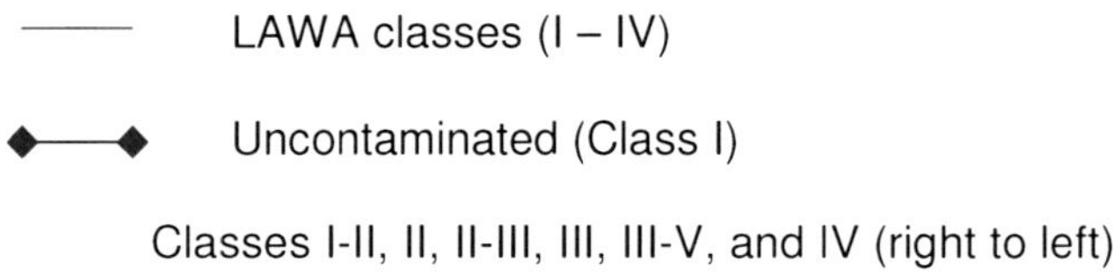

Fig. 5.8. Comparison of concentrations of Cu, Pb and Zn bound to SPM based on the LAWA classes, from the “Very strongly contaminated” left upper corner to the “Uncontaminated” lower right corner (groundwater).

Figure 5.7 and 5.8 show the LAWA classifying metals contamination of the SPM in surface water and groundwater. The obtained results show that 40% of the surface water samples are moderately contaminated class II with Cu and 28% moderately to strongly contaminated class II - III / III with Zn. Contamination due to Pb bound to SPM fall in the strongly contaminated class III in only 14 % of the surface water samples. According to the LAWA classification, in all water samples concentrations of Cu, Zn, and Pb bound to SPM are higher than the threshold value (class I) that indicates no contamination. The results also show that 70% of the groundwater samples are moderately to strongly contaminated classes II - III / III with Cu while 54% of the samples are strongly to very strongly contaminated classes III - IV with Zn. Around 46% of the groundwater samples are moderately to strongly contaminated classes II - III with Pb.

The calculated distribution coefficients (K_d) of heavy metals and As in the two-phase system (water-solid) reveal that the highest proportions of Cu, Zn, Ni, As and Pb are bound to the suspended particulate matter (Fig. 5.9a and 5.9b). The K_d values of Fe bound to SPM are the highest calculated. Since concentrations of Fe bound to SPM

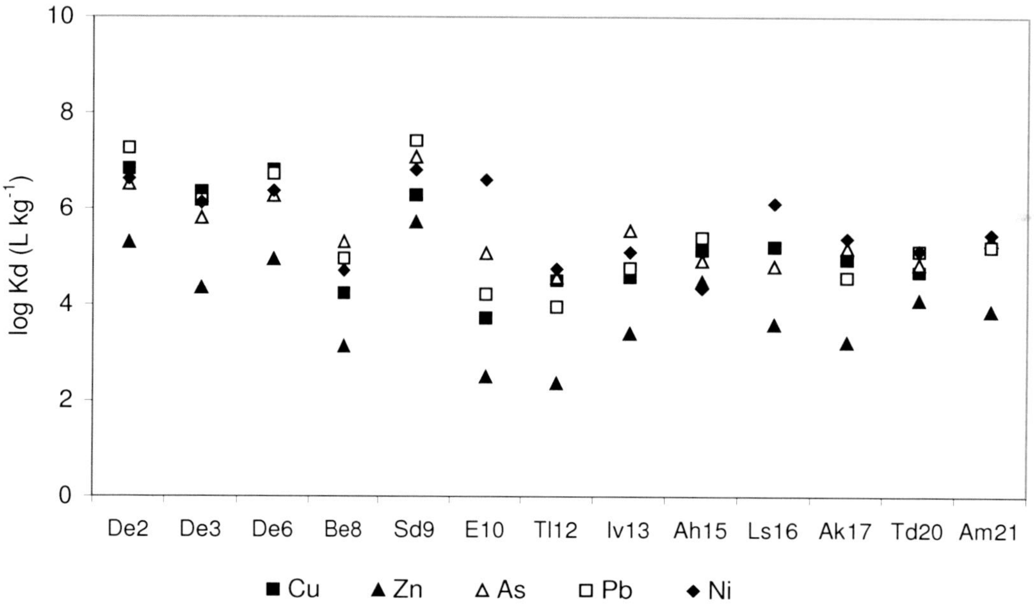

Fig. 5.9a. The distribution coefficients (K_d) of Cu, Zn, As, Pb and Ni tend to increase in samples taken close to the Aba Samuel Lake and the Akaki River (Groundwater).

are much higher than those of Cu, Zn, Ni, As and Pb, it is difficult to plot the Kd values for Fe in the same plot. In comparison with the other heavy metals and As the Kd values of Mn are also significantly higher.

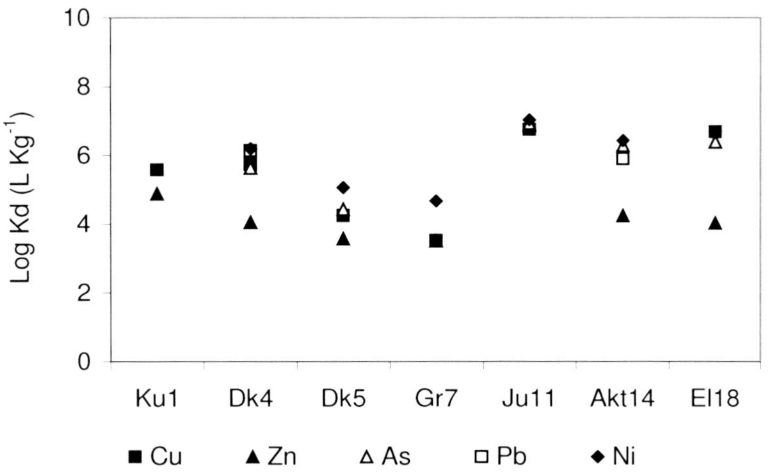

Fig. 5.9b. The distribution coefficients (K_d) of Cu, Zn, As, Pb and Ni (Surface water).

In surface water samples, the highest log Kd values 7.04 and 6.93 are determined for Ni and As respectively while the lowest K_d values 3.51 and 3.58 are calculated for Zn at sample sites Gr7 and Dk5. On the other hand, groundwater samples show highest K_d values for Pb (7.26) at De2 and Ni (6.81) at Sd9. Similar to surface water samples lowest K_d values are also measured in groundwater samples for Zn (2.50) at T12 and Cu (3.72) at E10. Comparison of Kd values between surface water samples (Fig. 5.9b) show that concentrations of Ni and Pb bound to SPM are the highest measured at the downstream part of the sampling area (e.g. E10). This sample site is a water well drilled very close to the Little Akaki River near Jugan at sample site Ju11. In groundwater, Ni increased from 50 µg/g at sample site Sd9 to higher concentrations of 4300 µg/g at sample site Tl12. At the Akaki well fields, concentrations of heavy metals bound to SPM are lower as compared to the upstream and downstream parts of the study area.

In water samples collected in 2003 suspended particulate matter (SPM) and several heavy metals are detected more commonly and in higher concentrations in Akaki Rivers and Aba Samuel Lake than at sites that are located upstream of the sub-basin. The highest concentrations of suspended particulate matter (5340 mg/l) are measured in the small Akaki River at Ju11 (Fig. 5.10). Other surface water samples at sites Gr7, Ku1, Dk4 and Dk5 also show high concentrations of suspended particles that reach 4000 mg/l, 3800 mg/l, 620 mg/l and 568 mg/l respectively. Considerable amounts of SPM are also measured in groundwater samples taken from water wells close to the Akaki Rivers and the Aba Samuel Lake. In groundwater, concentrations of SPM that are represented by the colloidal fractions are as high as 7 mg/l in samples taken from site Be8 (Fig. 5.10). Nearly the same amounts of suspended particulate matter/colloidal fractions are also measured in groundwater samples taken from sample sites at Ls16 and Ak17. Relatively the lowest amounts of SPM/colloidal matter are measured in groundwater sample at site E10. In other groundwater samples, concentrations of suspended particulate matter/colloidal fractions fall in the range between 3 to 5 mg/l.

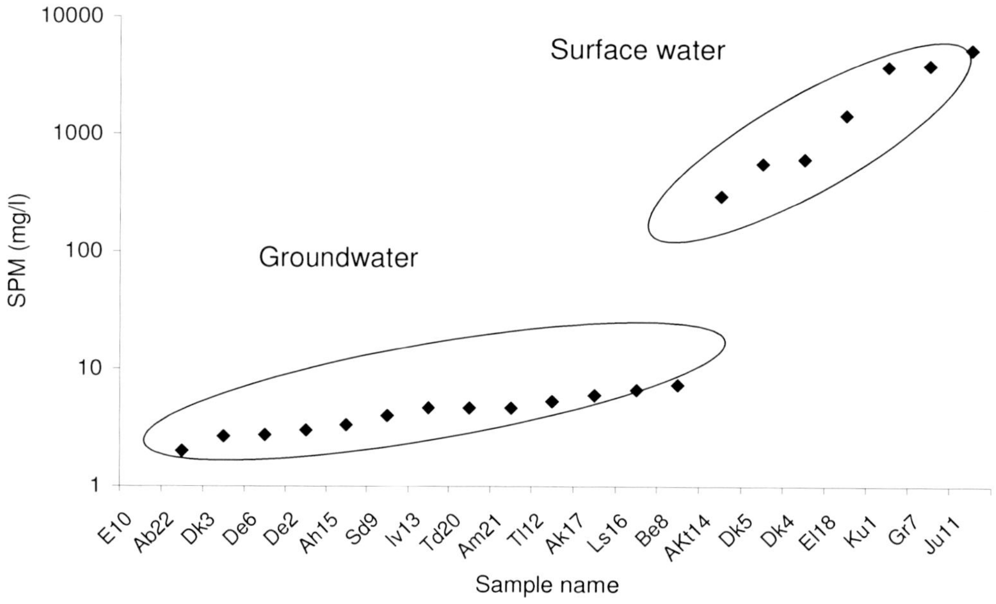

Fig. 5.10. Concentrations of SPM in surface water and groundwater samples (mg/l).

Results of the SEM analysis of suspended particulate matter samples from the Akaki River at sample site El18 are presented below. The SPM in this part of the Little Akaki River contains several fecal and clay sized particles (Fig. 5.12 and Fig. 5.13).

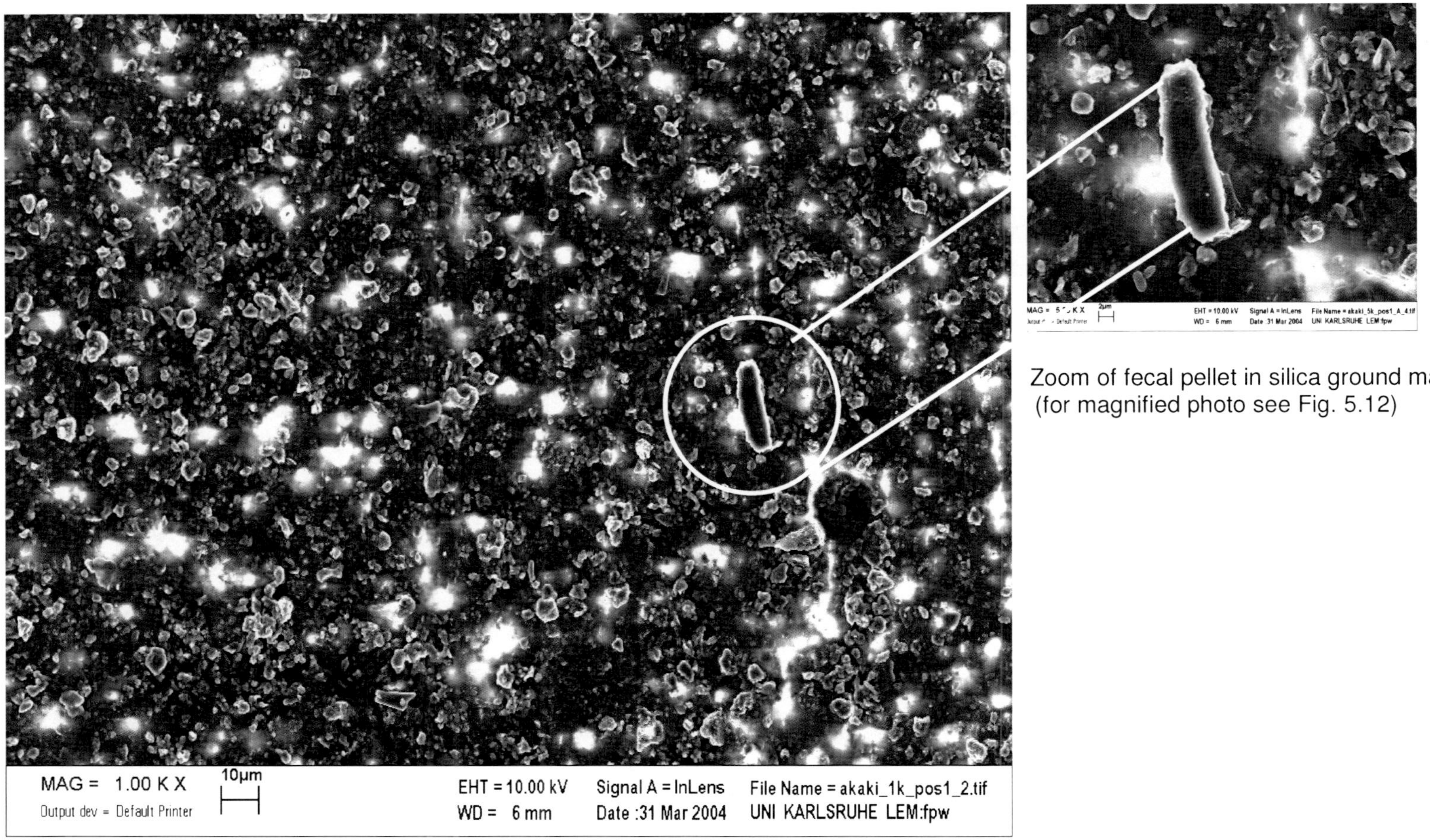

Zoom of fecal pellet in silica ground mass (for magnified photo see Fig. 5.12)

Fig. 5.11. SEM photo of the matrix of SPM taken from the Little Akaki River at sample site EI18 (ELICO).

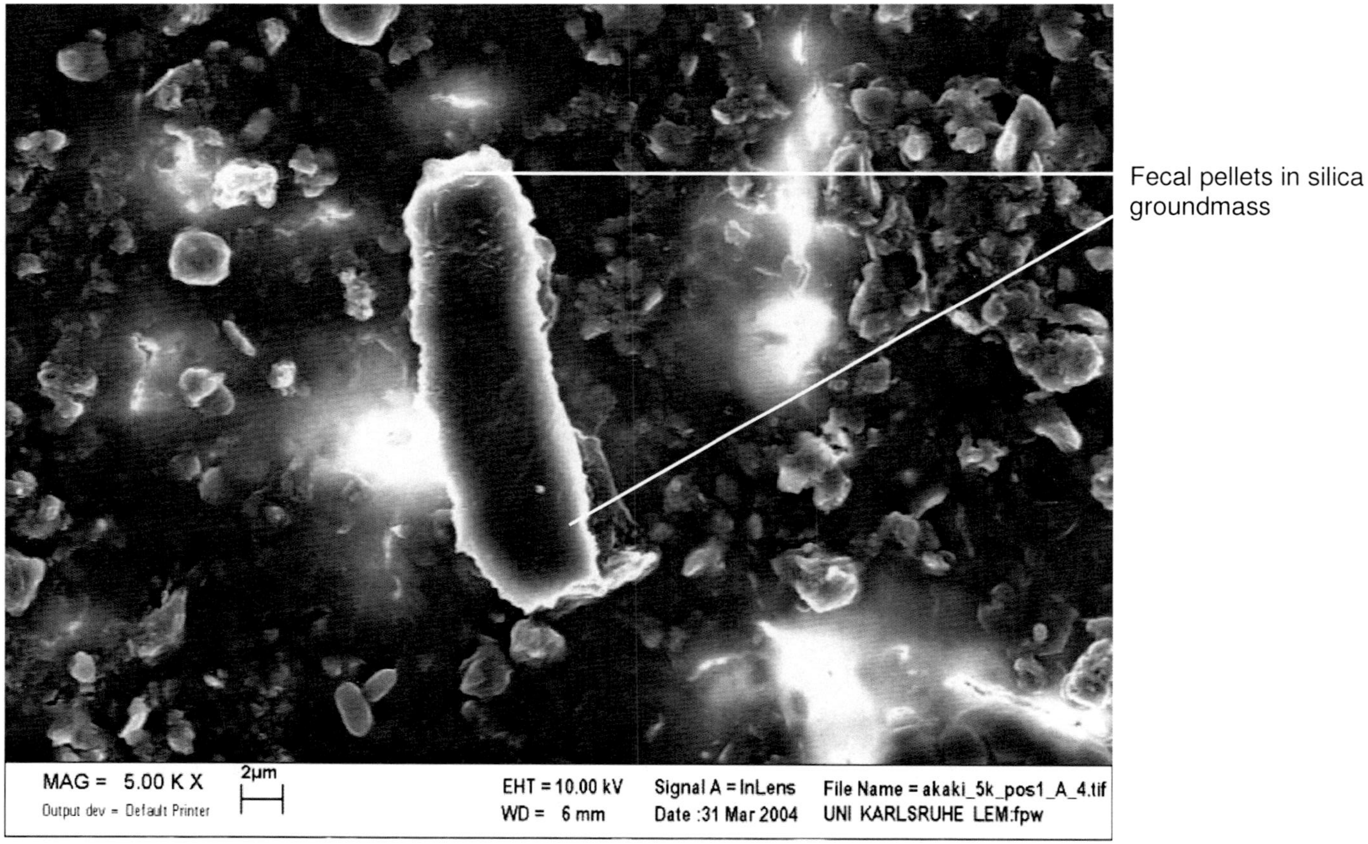

Fig. 5.12. SEM photo of SPM taken from the Little Akaki River at sample site EI18 (ELICO). Fecal pellets constitute part of organic particulate matter as seen in the middle of the photo.

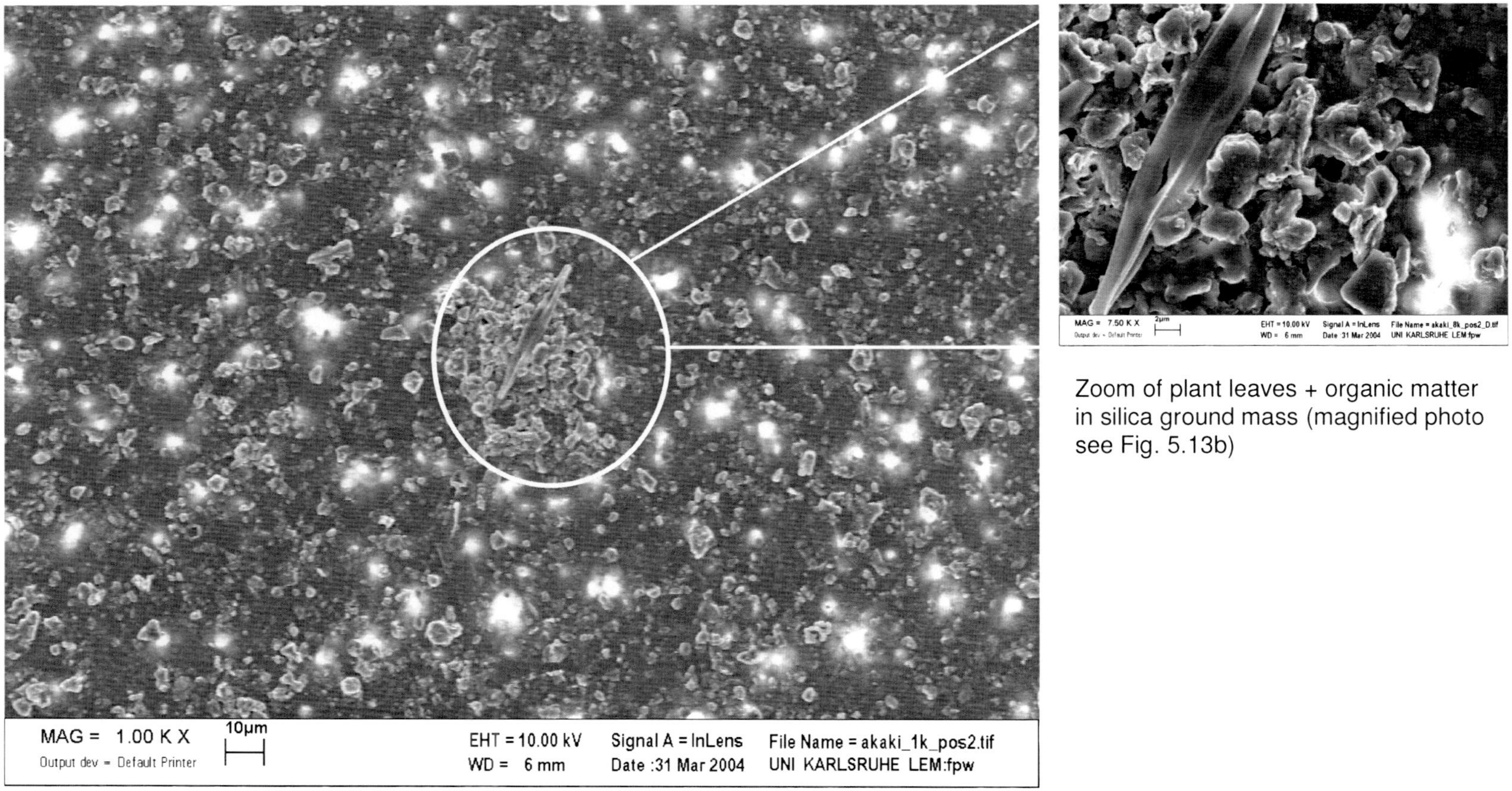

Fig. 5.13a. SEM photo of the matrix of SPM taken from the Little Akaki River at sample site EI18 (ELICO). Elongated plant leaves represent organic matter.

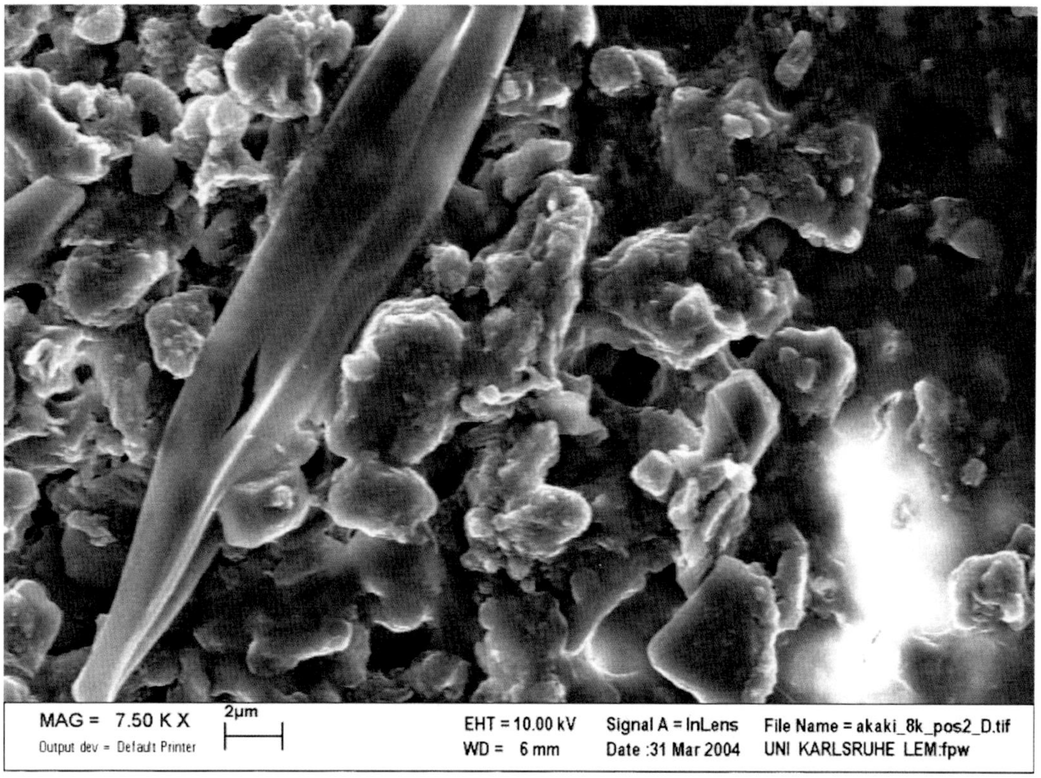

Fig. 5.13b. SEM photo of SPM taken from the Little Akaki River at sample site El18 (ELICO). Elongated plant leaves and other remains of organic matter represent organic fraction.

Organic matter and silicates constitute the major part of the suspended particle matter with few Fe oxide fractions. As determined by scanning electron microscopy the main compositions of SPM are clay minerals (montmorillonite), organic matter and some silt fractions. Fe oxides also constitute considerable amounts of the suspended particulate matter.

5.2 Characterization of Filter Media

5.2.1 Mineral Phases and Chemical Composition

The chemical composition and some physical properties of filter materials collected during the 2003 field trip are listed in table 5.5. Based on their chemical, mineralogical and hydrogeological parameters and their proximity to the target area three filter materials taken from Ethiopia are selected for in depth study of heavy metal retention capacities. These filter media are DQ (Dika) scoria, Kumo (KVA) volcanic ash, and Ethiopian Brick Factory (EBF) laterite. For similar investigation, two materials collected from Indonesia are also evaluated: Manganese rich (MnR) sediment and PYC (Merapi) pyroclastics.

It is clearly observed that silica and alumina are the main chemical compounds in almost all filter materials except in MnR sediment. Mineralogically KVA volcanic ash is mainly characterized by amorphous phases (75%). The type of volcanic glass is largely composed of rhyolitic, dacitic, and andesitic tephras. According to Yamada and Shoji (1983), morphology of volcanic glass varies depending on chemical composition, gas content of magma and type of volcanic eruption. Small amounts of quartz SiO_2 (4%), anorthoclase (3%), albite (8%) and clino pyroxene (5%) are present in KVA volcanic ash. A considerable amount of clay fraction (5%) is also measured.

Characteristic minerals in DQ scoria include albite (13%), clino pyroxene (5%), kaolin (12%), and hematite (10 - 20%). Quartz is present in minor amounts (7%). The main chemical compounds in DQ scoria are silica (46%) and alumina (16%). In addition to silica and alumina, other significant components that constitute DQ scoria include MgO (10%) and Fe_2O_3 (13%).

Similarly, the main chemical compounds in EBF laterite are also silica and alumina where the later is the highest in EBF laterite (Table 5.5). Chemical composition analyses reveal that EBF laterite consists of more than 57% SiO_2, 21% Al_2O_3 and around 13% Fe_2O_3. Mineralogically the material contains substantial amounts of quartz (57%) and albite (16%). Some clay minerals including kaolin (12%) and Montmorillonite (5%) also constitute significant fractions in EBF laterite. The physical and chemical properties, considerable specific surface area, and adsorptive capacity of red clay make it a potential material for metals removal from water/wastewater.

In MnR sediment, MnO (35%) is the dominant chemical component followed by SiO_2 and Al_2O_3 that compose 14% and 10% of the total oxide composition respectively.

Major oxide composition data also show that in addition to MnO, SiO_2 and Al_2O_3, other components such as CaO (9%) and Fe_2O_3 (8.4%) are also present in MnR sediment (Table 5.5). The main mineral phase that constitutes MnR sediment is birnessite (40%). Other fractions that represent the amorphous phase are mangano oxihydroxide phase.

Table 5.5. Chemical composition and physical property of the different filter media.

Material	KVA	EBF	DQ	MnR	PYC
Na_2O [%]	4.66	0.45	1.95	0.05	3.30
MgO [%]	1.84	1.01	9.57	0.26	2.72
Al_2O_3 [%]	13.32	21.13	15.69	10.40	18.91
SiO_2 [%]	68.32	57.81	45.54	13.63	54.10
P_2O_5 [%]	0.08	0.17	0.38	2.63	0.30
CaO [%]	1.26	0.41	8.58	9.18	10.60
K_2O [%]	3.86	1.91	0.58	0.5	2.50
TiO_2 [%]	0.70	2.17	1.99	1.21	0.74
MnO [%]	0.20	0.39	0.18	34.11	0.20
Fe_2O_3 [%]	5.26	13.10	12.58	8.38	7.30
Total [%]	99.5	98.6	97.0	80.4	100.7
Ni (ppm)	1.03	624.71	480.96	20	20.00
Cu (ppm)	10.95	56.97	53.90	54.65	43.27
Zn (ppm)	194.85	94.64	160.13	157.68	81.42
As (ppm)	7.17	5.00	8.83	14.84	5.43
Pb (ppm)	29.12	5.00	28.47	5	14.22
Density(g/cm^3)	0.67	0.85	1.29	1.12	0.39

Mineralogically, PYC pyroclastics is characterized by glassy texture. As in KVA volcanic ash, amorphous phases also represent the main mineralogy in PYC pyroclastics. Apart from amorphous phases, the main mineral phase includes anhortite. Results of chemical composition showed that SiO_2 (54%), Al_2O_3 (19%), CaO (11%) and Fe_2O_3 (7.3%) are the major oxides constituting PYC pyroclastics (Table 5.5).

The background heavy metal concentrations of all filter materials are also presented in table 5.5. It is possible to observe that concentrations of Ni, Cu, and Zn in all filter materials are the dominant ones. However, the mobile and acid exchangeable fractions of these heavy metals are well below the MCL of US EPA, which make them appropriate in the selection of filter media for water purification.

5.2.2 Granulometric Analysis and Hydrogeoloogical Parameters

The porosities of the materials are increasing as is written in the following order: MnR sediment, PYC pyroclastics, EBF laterite, DQ scoria and KVA volcanic ash. Hydrogeological parameters including bulk density, dry density, porosity and Hazen's permeability (k_f) of the filter materials are presented in table 5.6.

Table 5.6. Specific hydrological parameters of the filter materials.

Materials	Bulk Density (g/cm^3)	Water content %	Dry Density (g/cm^3)	Porosity %	k_f (m/s)
KVA	0.67	0.21	0.67	50.0	2.35E-05
DQ	1.35	4.85	1.29	40.0	1.86E-03
EBF	1.05	22.9	0.85	35.0	9.40E-05
MnR	1.13	1.31	1.12	22.7	1.04E-05
PYC	0.79	102	0.39	34.3	1.16E-04
QTZ	1.25	1.50	1.23	42.5	-

Except in MnR sediment, the porosities of the filter materials are above 30% as shown in table 5.6. The results of grain size distribution analysis show that the selected materials fall in the grain size range of coarse sand to silt, 1.0 - 0.049 mm according to the classification by USDA soil textural classification system (Table 5.7). Based on this classification the coarsest material investigated is Dika scoria (1.0 - 1.99 mm) that corresponds to very coarse sand. The fine grain materials analyzed range from 0.1 mm to 0.049 mm (KVA volcanic ash) that belongs to the silt-clay fraction.

Table 5.7. Classification of the filter materials based on grain size distribution. (USDA soil textural classification system, Brown 2003).

Name	Diameter limits	DQ scoria	EBF laterite	PYC pyroclasts	KVA vol ash	MnR sediment
	mm	Percentage passing (%)				
Very coarse sand	1.6	66.58	23.74	21.34	7.98	26.69
Very coarse sand	1	9.97	19.32	15.73	6.17	26.35
Coarse sand	0.5	11.32	23.54	22.24	11.05	23.82
Medium sand	0.25	6.20	13.68	15.19	13.55	9.12
Fine sand	0.125	4.85	7.44	13.02	21.99	3.72
Very fine sand	0.05	1.08	5.23	9.04	23.49	3.89
Silt	0.05-0002	0.00	3.82	6.33	21.39	10.14

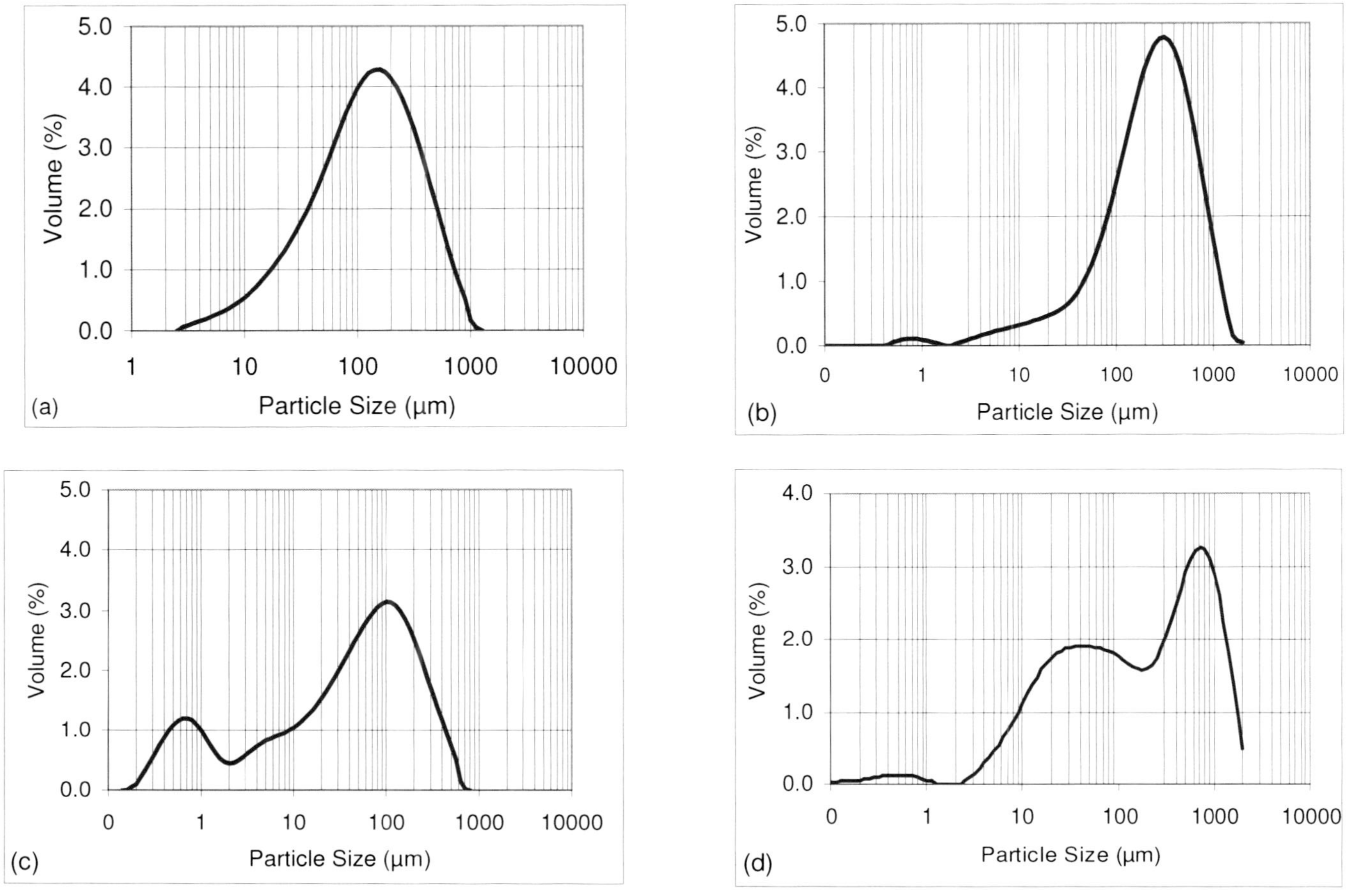

Fig. 5.14. The percentages of grain size fractions in (a) KVA volcanic ash, (b) DQ scoria, (c) EBF laterite, (d) MnR sediment and (e) PYC pyroclastics.

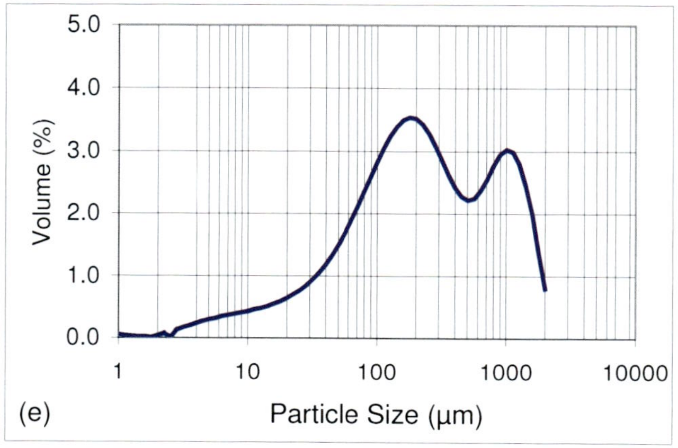

Fig. 5.14. The percentages of grain size fractions in (a) KVA volcanic ash, (b) DQ scoria, (c) EBF laterite, (d) MnR sediment and (e) PYC pyroclastics (*cntd.*).

Key based on grain size classification (grain size in mm)

German Standard	Boulders and Cobbles	Gravel	Sand	Silt	Clay
(DIN 4022)	>63	63 - 2.0	2.0 - 0.06	0.06 - 0.002	< 0.002

According to results of grain size analysis the materials show wide range of grain sizes and are given in figure 5.14. Furthermore based on the DIN 4022 classification, KVA volcanic ash fall in to silt/sand fraction while DQ scoria belongs to the fine sand fraction. EBF laterite shows all size ranges from clay to silt sand. MnR sediment and PYC pyroclastics also fall in to the silt/ sand fraction with minor clay fractions (Fig. 5.14).

5.2.3 Sequential Extraction

The results of sequential extraction scheme according to Zeien and Brümmer (1989) of the filter media are shown in figure 5.15. Results of KVA volcanic ash, DQ scoria and EBF laterite are determined based on this sequential extraction scheme. On the other hand, results of sequential extraction referred to MnR sediment and PYC pyroclastics are based on the BCR Three Step sequential extraction (Davidson et al. 1999, 2000). The sequential extractions used in this study are useful to indirectly assess the potential mobility and bioavailability of heavy metals once the filter media are used and disposed to the environment. The six chemical fractions used in Zeien and Brümmer (1989) are operationally defined by an extraction sequence that follows the order of decreasing solubility (Clevenger and Mullins 1982, Soon and Bates 1982, Lake et al. 1984, Xian 1989). Assuming that bioavailability is related to solubility, then metal bioavailability decreases in the order: water-soluble > exchangeable > carbonate > Fe-Mn oxide > organic > residual.

The results in both extraction schemes indicate that the sum of mobile and acid exchangeable fractions of heavy metal is well below the US EPA MCL. Though, the highest total concentrations measured after carrying out sequential extractions according to Zeien and Brümmer (1989) are well below the MCL higher concentrations of heavy metals are associated to the crystalline and non-crystalline Fe-oxide fractions as shown in figure 5.15. The heavy metal fractions on the crystallized Fe-oxide phase exceed that on the other fractions. However, heavy metals associated with Mn-oxide, organic, amorphous Fe-oxide, and crystalline Fe-oxide phases are less mobile and available than the water-soluble and exchangeable phases. The results are in agreement with other sequential extraction schemes performed on sediments and tailings. From these results, it is evident that the filter materials are suitable to water purification.

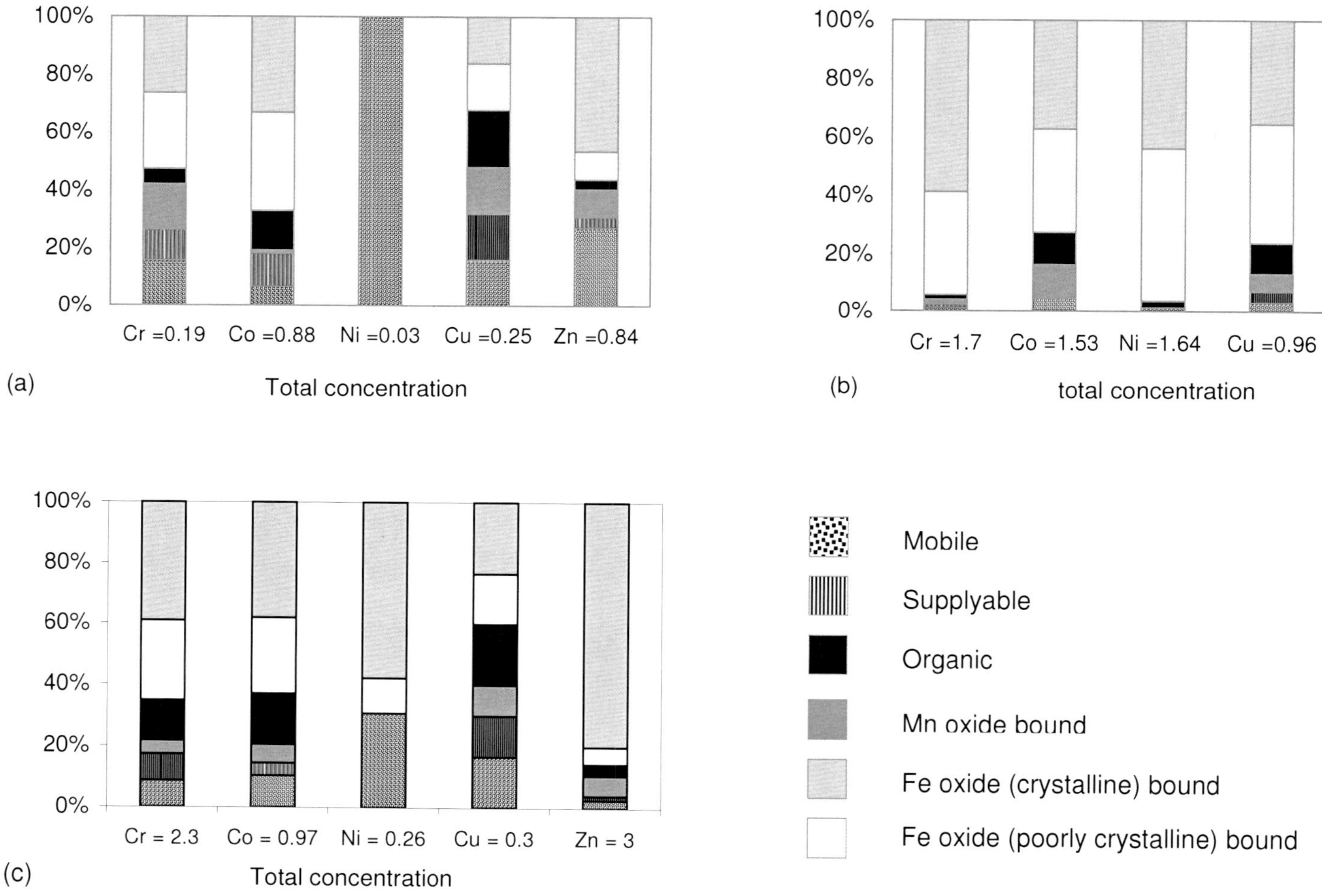
100%
80%
60%
40%
20%
0%
Cr =0.19
Co =0.88
Ni =0.03
Cu =0.25
Zn =0.84
(a)
Total concentration
100%
80%
60%
40%
20%
0%
Cr =1.7
Co =1.53
Ni =1.64
Cu =0.96
Zn =1.07
(b)
total concentration
100%
80%
60%
40%
20%
0%
Cr = 2.3
Co = 0.97
Ni = 0.26
Cu = 0.3
Zn = 3
(c)
Total concentration
Mobile
Supplyable
Organic
Mn oxide bound
Fe oxide (crystalline) bound
Fe oxide (poorly crystalline) bound

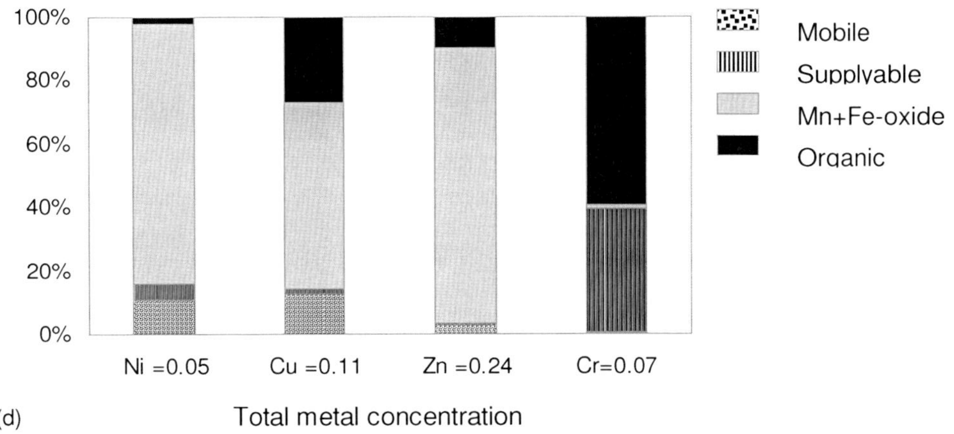

(d)

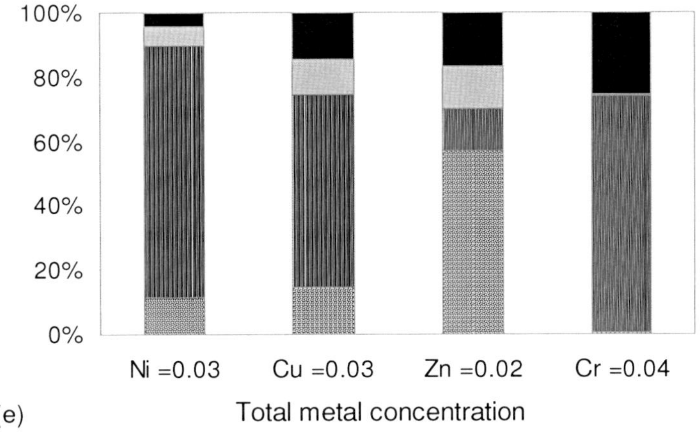

(e)

Fig. 5.15. The distribution of Cr, Co, Ni, Zn, and Cu in different fractions of (a) KVA volcanic ash, (b) DQ scoria, and (c) EBF laterite as were evaluated by sequential extraction after Zeien and Brümmer 1989. The evaluation on (d) MnR sediment and (e) PYC pyroclastics were done after extraction with BCR procedures (Davidson et al. 1999, 2000). Total metal concentrations are in mg/l.

According to the BCR Three Step sequential extraction higher proportion of heavy metals are associated to the Fe/Mn oxide fractions of MnR sediment. On the other hand, association of heavy metals in PYC pyroclastics is more pronounced in the acid exchangeable fractions than the other filter materials. Substantial amount of heavy metals are released from the acid exchangeable fraction of PYC pyroclastics.

5.3 Results of Column Experiment

5.3.1 Hydrochemical Compositions of Filter Effluents

5.3.1.1 pH, Dissolved Oxygen, Conductivity

The pH of the effluent solution shows a significant increase after passing through the columns regardless of the type of media packed in the columns; the pH change is from 7.0 of initial value to a final value of 8.5 in effluents of all columns (See Appendix). The highest effluent pH value is measured in the column filled with DQ scoria. In this effluent the pH remains above 8.4 throughout the experiment. The conductivity of the solution also increase slightly from an initial value of 184 µS/cm to a maximum of 226 µS/cm in the column packed with KVA volcanic ash showing an increase in the amount of dissolved ions. Almost in all of the effluents the concentrations of dissolved oxygen ranges between a minimum of 3 mg/l and a maximum of 8 mg/l. The lowest dissolved oxygen concentrations are measured in the effluents of KVA volcanic ash (2.5 mg/l) and of QTZ sand (2.3 mg/l) at day 42.

5.3.1.2 Anions Cl^-, SO_4^{2-} and NO_3^-

The anions concentrations of effluents show variable trends among the various filter media. Concentrations of Cl^-, SO_4^{2-} and NO_3^- measured during the 12 weeks experimental period are shown (Fig. 5.16). In all columns effluent concentrations of NO_3^- are almost below the detection limit. However, some high concentrations of NO_3^- are measured in columns filled with QTZ sand (18.7 mg/l) at day 36. There is a similar trend observed in NO_3^- concentrations of effluents over the experimental period. Between day 10 and 11 nitrate concentrations reach up to 10 mg/l in all columns tested. Some erratic nitrate concentrations are, however, observed in PYC pyroclastics at days 11, 15, 44 and 71 (8.9 mg/l, 7.7 mg/l, 10.5 mg/l, and 10.5 mg/l respectively). In columns filled with KVA volcanic ash, PYC pyroclastics, MnR sediment, DQ scoria, and EBF laterite 8.9 mg/l, 11.9 mg/l, 10.7 mg/l, and 7.47 mg/l NO_3^- respectively is measured in the effluents at day 11.

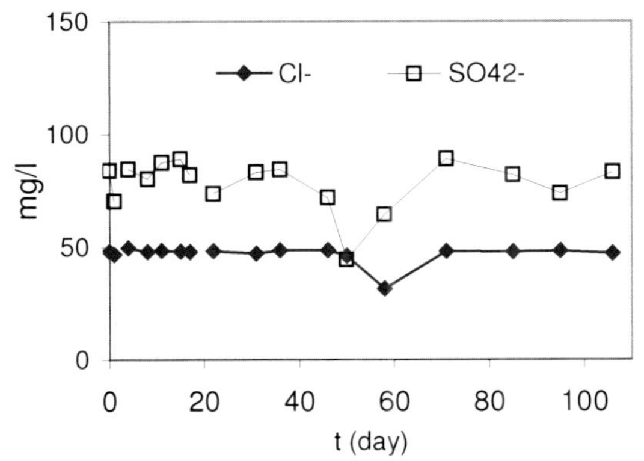

(a)

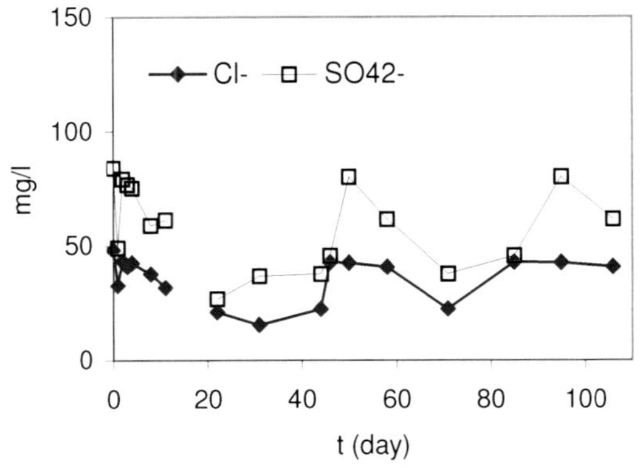

(b)

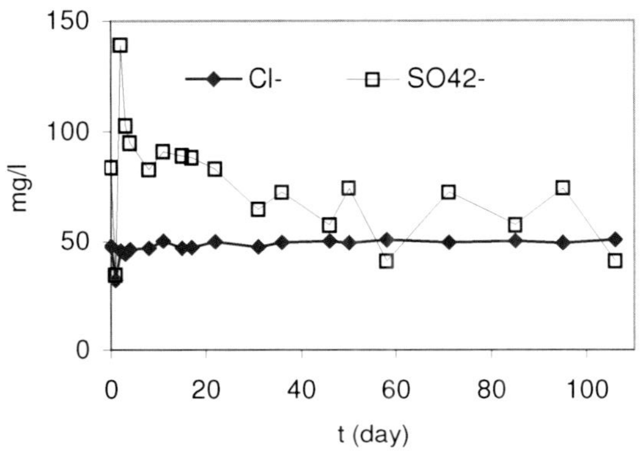

(c)

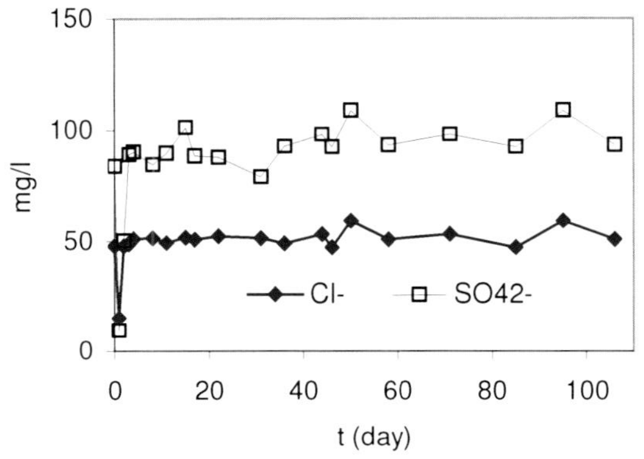

(d)

Fig. 5.16. The Cl^- and SO_4^{2-} concentrations of effluents in (a) KVA volcanic ash, (b) DQ scoria, (c) EBF laterite, (d) MnR sediment, (e) PYC pyroclastics and (f) QTZ sand. NO_3^- concentrations are below the detection limits in all columns.

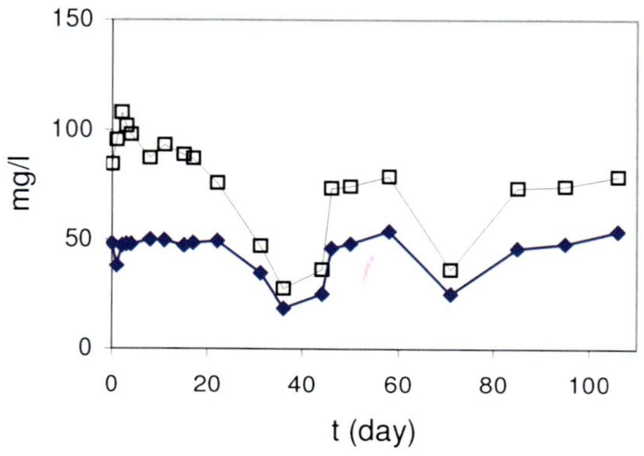

(e)

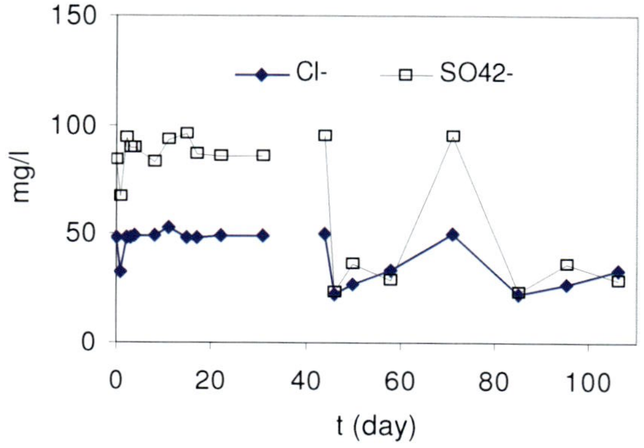

(f)

Fig. 5.16. The Cl^- and SO_4^{2-} concentrations of effluents in (a) KVA volcanic ash, (b) DQ scoria, (c) EBF laterite, (d) MnR sediment, (e) PYC pyroclastics and (f) QTZ sand. NO_3^- concentrations are below the detection limits in all columns *(contd.)*.

5.3.1.3 Dissolved Heavy Metal and Arsenic Concentrations

Heavy metals Zn, Cu, Ni, and Pb along with As are given emphasis in this column test while these metals are the dominant pollutants in the Akaki waters both in the dissolved and SPM bound phases. Comparison of the equilibrium and initial metal concentration ratios (C/C_0) of Zn, Cu, Ni, and Pb in all column media indicates that except Ni all metals show strong retention. Zinc, Cu and Pb are found to exhibit higher adsorption by all media where the maximum concentration ratios are lower than 0.1 for Zn while 0.2 for Cu and Pb. In column packed with MnR sediment the C/C_0 of Zn, Cu, Ni, As and Pb are the lowest observed (Fig. 5.17d) implying that metals retention is the highest. In this column all the heavy metals tested are retained well above 95% through out the experimental period. On the other hand, the highest C/C_0 ratios are observed in PYC pyroclastics for Ni (3.4), Cu (0.3), and Zn (0.03) followed by QTZ sand where the ratios are 0.74, 0.16, 0.08, and 0.33 for Ni, Cu, Zn, and Pb respectively (Fig. 5.17e and 5.17f). Removal of Ni, however, is variable based on the type of material filled in the columns. Similar trend of Ni removal is observed in columns packed with DQ scoria and EBF laterite where $C/C_0 < 0.15$ and < 0.12 respectively (Fig. 5.17b and 5.17c). On the other hand, Ni removal is relatively reduced in columns filled with KVA volcanic ash, PYC pyroclastics and QTZ sand (Fig. 5.17). The least performance in removing Ni from the effluents is observed in columns packed with PYC pyroclastics and QTZ sand.

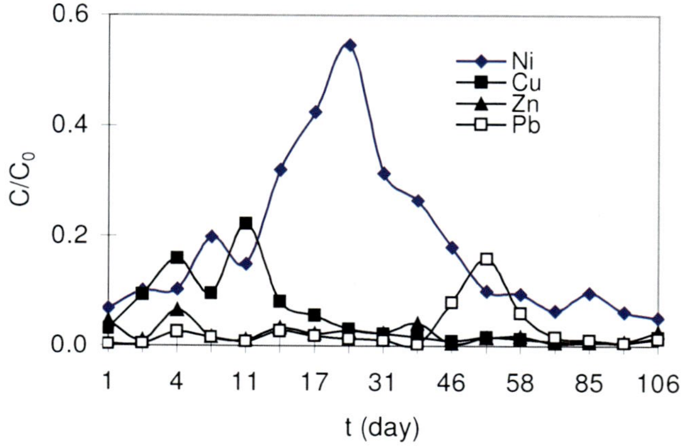

(a)

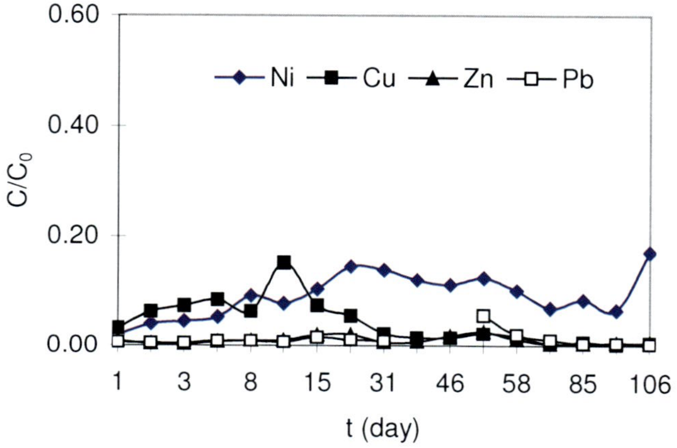

(b)

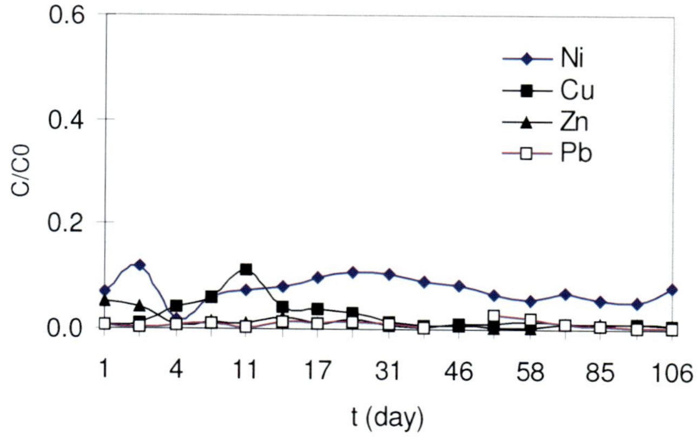

(c)

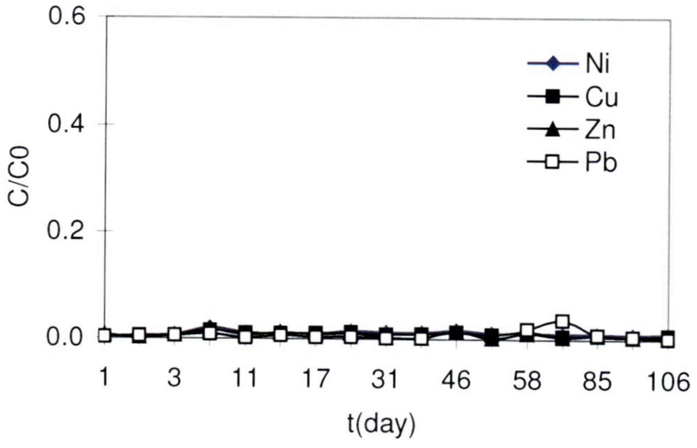

(d)

Fig. 5.17. C/C_0 of Ni, Cu, Zn, and Pb in (a) KVA Volcanic ash, (b) DQ scoria, (c) EBF laterite (d) MnR sediment (e) PYC pyroclastics, and (f) QTZ sand.

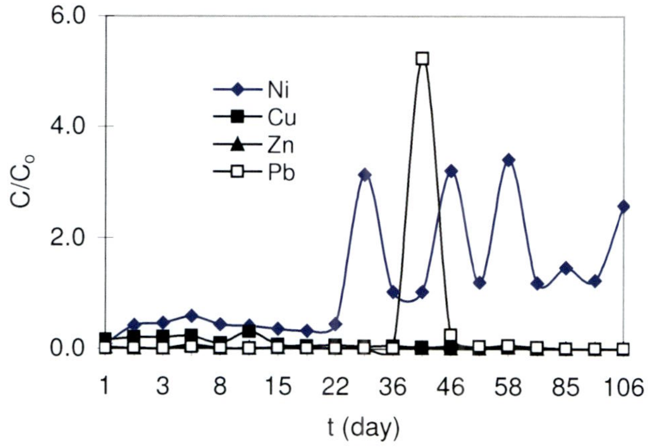

(e)

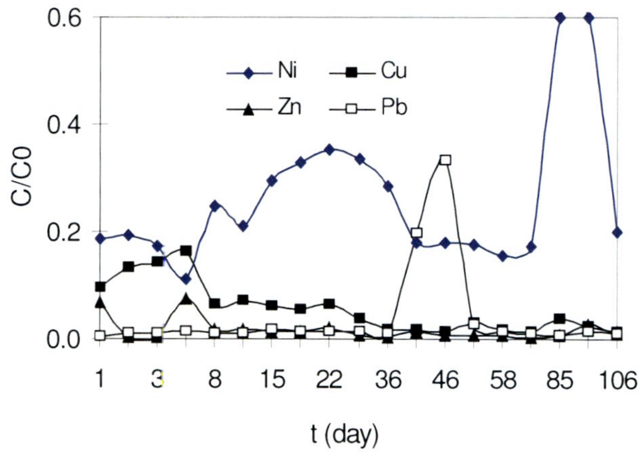

(f)

Fig. 5.17. C/C_0 of Ni, Cu, Zn, and Pb in (a) KVA Volcanic ash, (b) DQ scoria, (c) EBF laterite (d) MnR sediment (e) PYC pyroclastics, and (f) QTZ sand *(contd.)*.

The performance of the column media in reducing SPM and heavy metal both in the dissolved phase as well as bound to SPM from the synthetic solutions is assessed by averaging the removal efficiencies of the columns media over the experimental period of 12 weeks. The following relation is used to determine the removal efficiencies of the filter media towards each metal tested.

$$\text{Metal removal (\%)} = \frac{C_0 - C_w}{C_0} * 100$$

where C_0 = initial dissolved metal concentration

C_w = dissolved metal concentration after passing through the column media

The MnR sediment is found to be the best performing media in removing almost all of the heavy metal from the synthetic polluted water. Considering the filter media collected from Ethiopia the EBF laterite and DQ scoria are found to be better performing media in removing almost all of the heavy metal except As. More than 90% (mean value, n=17) of Ni, Cu, Zn and Pb are retained in columns filled with MnR sediment, EBF laterite and DQ scoria in a decreasing order respectively. The comparison among the removal efficiencies of all filter media in figure 5.18 and 5.19 indicates that EBF laterite (>90%) and DQ scoria (>90%) are relatively better performing media than the volcanic ash in removing As and Ni from the synthetic polluted water. In comparison to EBF laterite and DQ scoria (90%), the removal of Ni by KVA volcanic ash (80%) is slightly lower (Fig. 5.19). However, considerable amounts of Cu, Zn, and Pb (>90%, mean value, n=17) are removed by KVA volcanic ash, PYC pyroclastics, and QTZ sand (Fig. 5.19). The ability to reduce the levels of Cu, Zn and Pb do not show any significant variation among the filter media. For comparison purposes a selectivity series can be given for EBF laterite and DQ scoria as follows:

Zn > Cu > Ni > Pb > As.

Comparison of the apparent heavy metal removal efficiencies of the filter materials is also shown in (Fig. 5.18). The apparent removal efficiencies represent the removal capacities of the filter materials without the contribution of Na-montmorillonite and organic material spiked. This was determined after subtracting the heavy metal removal capacity of Na - montmorillonite and organic mater spiked from the removal efficiencies of the bulk column media. The apparent removal capacities of KVA volcanic ash, PYC pyroclastics and QTZ sand for Cu and Zn reach up to 44% and 71% respectively. DQ scoria and EBF laterite remove 73% of Zn from the solution while the same media removed only 48% of Cu (Fig. 5.18). Relatively higher retention of Ni is observed in DQ scoria (55%) and EBF laterite (56%) compared to QTZ sand that retains 38% of the total

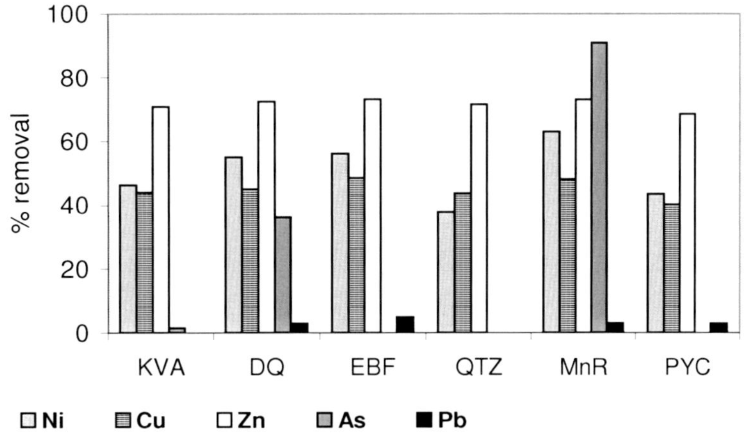

Fig. 5.18. Apparent removal efficiencies (%) of the filter material (removal capacities exclude contribution of SPM + organic matter).

Ni in the solution. The apparent removal capacity of KVA volcanic ash for Ni also reaches 46%. MnR sediment retains significant amounts of Ni (63%), Cu (48%), and Zn (73%) from the influents. Apparent removal of As by the filter media is only effective in columns packed with DQ scoria (36%) and MnR sediment (90%). Arsenic removal by the remaining filter materials is almost negligible. The apparent removal capacities of all filter media for Pb are the lowest calculated indicating that removal of Pb is almost entirely attributed to the spiked SPM and organic matter in the synthetic polluted water (Fig. 5.19).

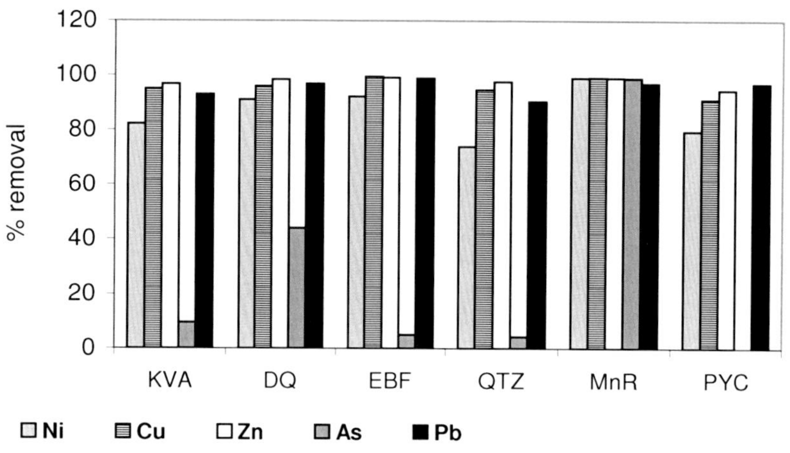

Fig. 5.19. Removal efficiencies (%) of the bulk filter media (including the removal capacities of SPM and organic matter).

Another interesting phenomenon occurred during the column experiment is the release of Fe and Mn from the columns. In spite of the fact that the initial concentrations of Fe and Mn are low in the simulated polluted water release of these metals were observed during the experiment (Fig. 5.20a - 5.20f). Iron and Mn are not spiked in the simulated polluted water and have initial concentrations of 0.65 µmol/l and 0.21 µmol/l respectively. All concentrations are mean values for n=17 samples. In the effluents of EBF laterite 22.3 µmol/l of Mn is measured while MnR sediment shows the lowest concentration (0.5 µmol/l). The released Mn is also higher in the effluents of DQ scoria (11.6 µmol/l) and KVA volcanic ash (6.8 µmol/l). Significant amounts of Mn 1.2 µmol/l and 3.4 µmol/l are released from columns filled with QTZ sand and PYC pyroclastics respectively.

As compared to Mn, Fe is released at lower concentrations from columns filled with DQ scoria (0.51 µmol/l) and EBF laterite (0.53 µg/l). Nevertheless, the highest release of Fe is measured from columns filled with PYC pyroclastics (0.83 µmol/l) and KVA volcanic ash (0.55 µmol/l) (Fig. 5.20a). The lowest concentrations of Fe 0.03 µmol/l and 0.4 µmol/l are released, however, from columns filled with MnR sediment and QTZ sand respectively (Fig.5.20b and 5.20c).

Arsenic is also released from almost all columns over the experimental period. High variations in effluent concentrations of As from columns filled with KVA volcanic ash, DQ scoria, EBF laterite and QTZ sand are shown that are attributed to desorption from the respective columns. However, the highest released average concentration of As (14.5 µmol/l) is observed in effluents of KVA volcanic ash and QTZ sand. The later shows release of As that reach 16.1 µmol/l (Fig. 5.20d). Similar As release is also observed from columns filled with DQ scoria and EBF laterite (8.93 µmol/l and 5.72 µmol/l) respectively (Fig. 5.20a). On the other hand, the lowest release of As is observed from columns filled with MnR sediment (0.18 µmol/l).

Though Cd is not spiked in the initial simulated polluted water, small amounts of Cd are released from all columns. The average released concentrations are determined for effluent samples collected during the overall experiment (n=17). The lowest desorbed concentrations of Cd are found in the effluents of MnR sediments (0.001 µmol/l) while the highest are in PYC pyroclastics (0.01 µmol/l). In effluents of KVA volcanic ash and QTZ sand desorbed concentrations reach 0.07 µmol/l and 0.006 µmol/l Cd respectively.

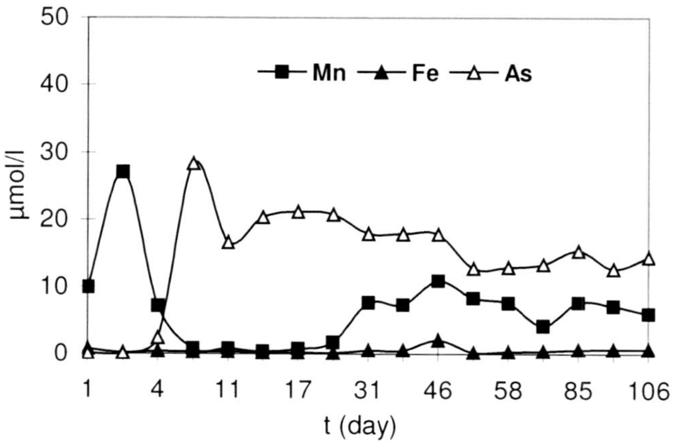
Mn
Fe
As
µmol/l
50
40
30
20
10
0
1
4
11
17
31
46
58
85
106
t (day)

(a)

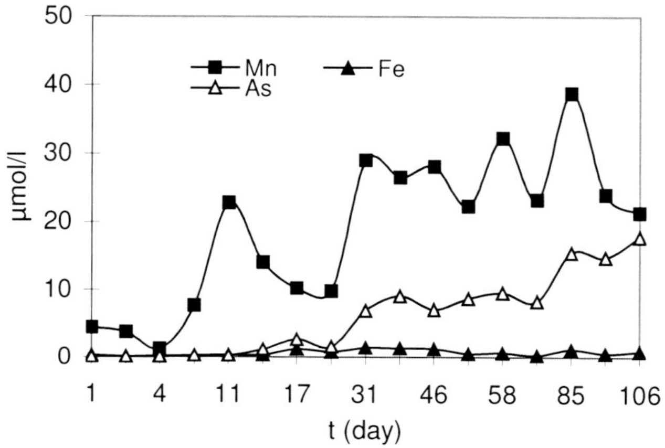
Mn
Fe
As
µmol/l
50
40
30
20
10
0
1
4
11
17
31
46
58
85
106
t (day)

(b)

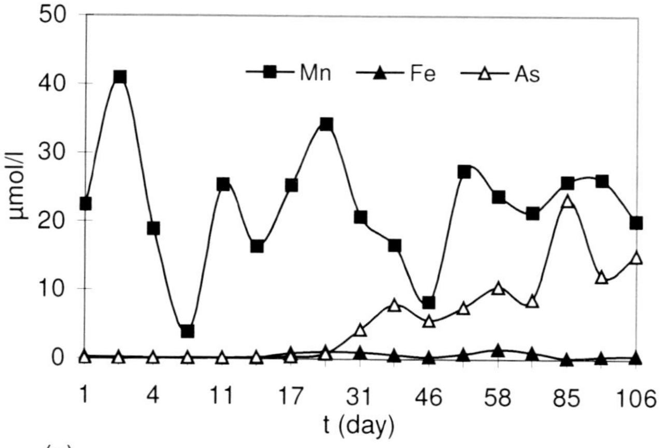
Mn
Fe
As
µmol/l
50
40
30
20
10
0
1
4
11
17
31
46
58
85
106
t (day)

(c)

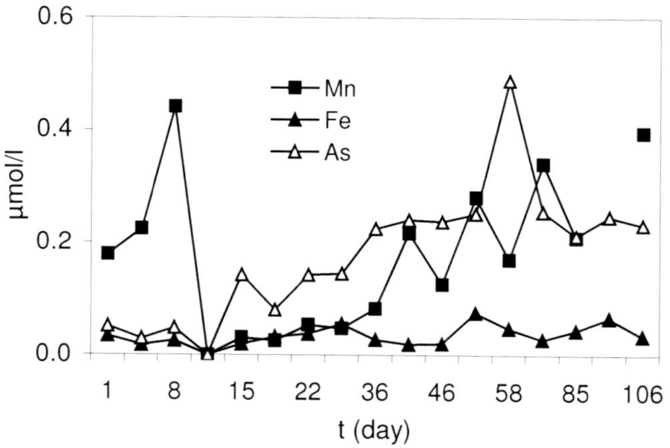
Mn
Fe
As
µmol/l
0.6
0.4
0.2
0.0
1
8
15
22
36
46
58
85
106
t (day)

d)

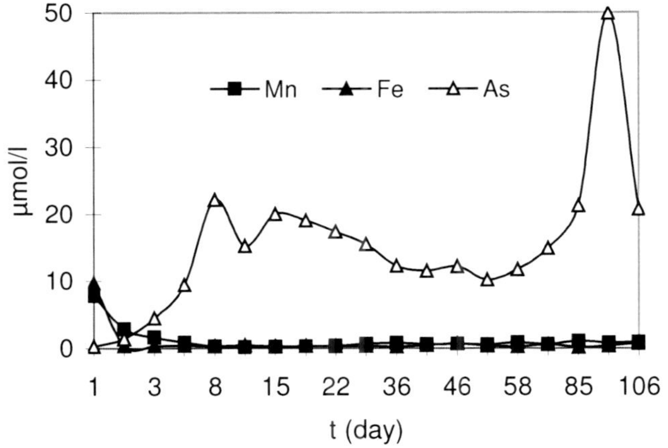

(e)

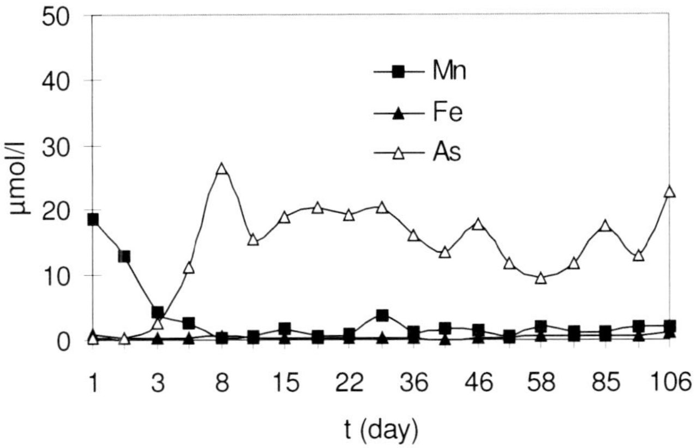

(f)

Fig. 5.20. The release of Mn, Fe, and As from a) KVA volcanic ash, b) DQ scoria, c) EBF laterite, d) MnR sediment, e) QTZ sand and f) PYC pyroclastics. (Initial solution concentrations of Fe=0.65 µmol/l; Mn=0.21 µmol/l, and As=13.34 µmol/l).

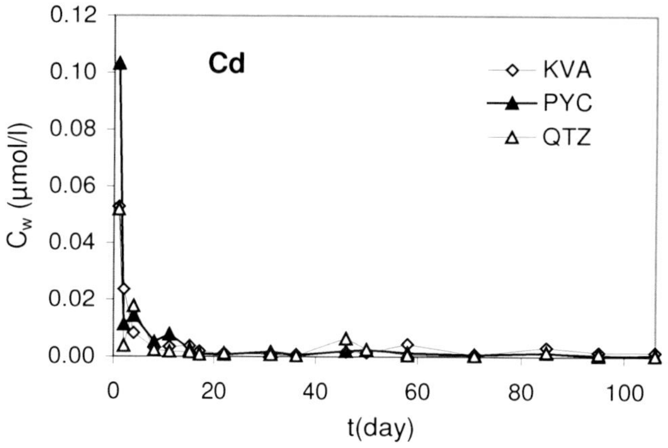

(a)

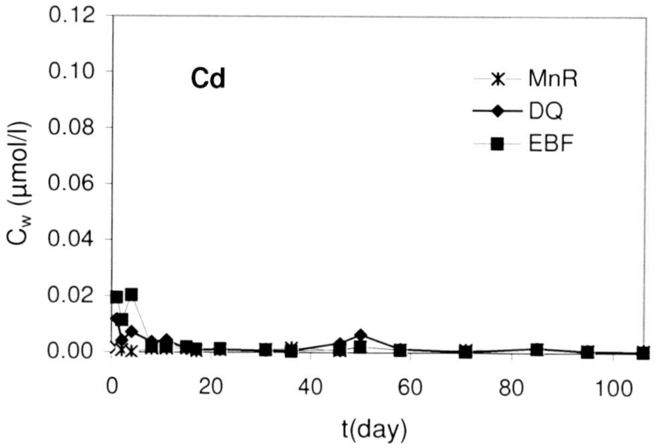

(b)

Fig. 5.21. The release of Cd from (a) KVA volcanic ash, PYC pyroclastics, and QTZ sand (b) MnR sediment, EBF laterite, and DQ scoria (Initial solution concentrations of Cd=0.02 µmol/l).

Concentration of Cd up to 0.0039 µmol/l is measured in the effluents of EBF laterite while DQ scoria show similar concentration (0.0029 µmol/l) (Fig. 5.21a and 5.21b).

5.3.2 Removal of SPM and Heavy Metal Bound to SPM

Heavy metals analysis on filtered suspended particulate matter (>0.45 µm) from the Akaki River and groundwater samples show that up to 74 mg/g Fe is bound to SPM of the Akaki River. The concentrations of Mn, Zn, Cu, Ni, and Pb bound to SPM in surface water samples exceed concentrations that are expected in natural unpolluted surface water (Tolera et al. 2006). On the other hand, groundwater samples also contain substantial amounts of Fe, Mn, Ni, Cu, Zn, As, and Pb bound to SPM. High concentrations of Fe (138 mg/g), Mn (5.9 mg/g), Cu (5.8 mg/g), Zn (37.6 mg/g), As (10.6 mg/g), Pb (6.2 mg/g), and Ni (4.4 mg/g) are detected in SPM fractions of groundwater used for drinking.

Distribution coefficient (K_d) values of heavy metals in surface water and groundwater samples also supported the high association of heavy metals with SPM. The K_d values of As, Ni and Pb determined considering the dissolved and SPM phases of heavy metal reach up to 3000 l/g, 4000 l/g, and 18,000 l/g in groundwater samples respectively. In surface waters values up to 428 l/g, 1567 l/g, and 1380 l/g are calculated for As, Ni, and Pb respectively (Tolera et al. 2005). The amounts of SPM are reduced to a maximum of 100 mg/l from an initial concentration of 1000 mg/l by all columns tested (Fig. 5.22). All columns remove more than 90% of the spiked SPM through out the experiment period. The filter media also retained metals that have been already sorbed on the suspended particulate matter before passing through the column. In columns filled with PYC pyroclastics and EBF laterite the effluent concentrations of SPM seem to remain constant through out the experiment. Similar trend in the reduction of the amounts of SPM by KVA volcanic ash and QTZ sand is shown. In general the lowest concentrations of SPM are measured in the effluents taken from the MnR sediments.

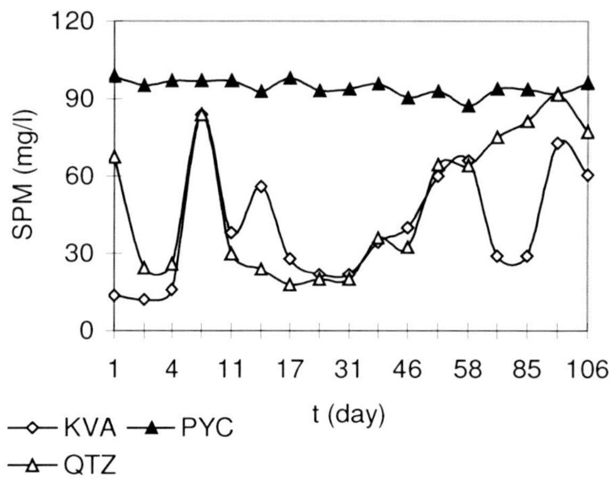

(a)

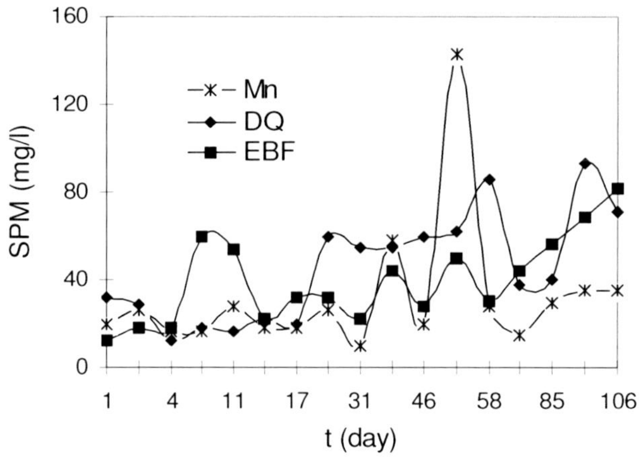

(b)

Fig. 5.22. Effluent concentrations of SPM in column experiment (a) KVA volcanic ash, PYC pyroclastics, and QTZ sand; (b) MnR sediment, DQ scoria, and EBF laterite. The initial SPM concentration in the synthetic wastewater is 1000 mg/l.

However, early clogging of the filter media by SPM was inevitable in columns packed with KVA volcanic ash and EBF laterite. The amount of SPM within the columns increased with time as was evidenced by the decreased amount of SPM in the effluents. Filter media with higher contents of silt and clay fractions might cause clogging earlier than those such as QTZ sand. This is the case for EBF laterite and KVA volcanic ash.

Saturation of columns with SPM is noticed at day 44 in columns filled with KVA volcanic ash, DQ scoria, and EBF laterite. Earlier clogging of the media is even observed in column filled with EBF laterite that has considerable amounts of kaolin and montmorillonite.

5.3.3 Sequential Extraction and Leaching of Heavy Metal

Single and sequential extraction procedures performed on the used filter media reveal the distribution of heavy metal in the columns sludges. Certain possible removal/release mechanisms of heavy metals associated with the used filter media are speculated. The water dissolved fractions of Ni, Cu, Zn, As, and Pb are compared in figure 5.23. Results of single extraction of the used media display that the heavy metals of interest are less likely adsorbed in the water-soluble fraction.

However, compared to the other heavy metals the water-soluble fraction of As is higher in columns packed with KVA volcanic ash, DQ scoria, EBF laterite and QTZ sand (65 µg/g, 72 µg/g, 45 µg/g and 33 µg/g respectively). This indicates that As is weakly bound to the surfaces of these filter media. Nickel is another metal showing relatively higher concentrations in the water-soluble fraction. From columns packed with KVA volcanic ash, DQ scoria, EBF laterite and QTZ sand 60 µg/g, 44 µg/g, and 33 µg/g and 46 µg/g of Ni are released after leaching the used media with water. The releases of Cu from the same columns reach 20 µg/g, 27 µg/g, 21µg/g and 17 µg/g respectively. These concentrations are even lower for Zn (37 µg/g, 12 µg/g, 5 µg/g and 22 µg/g respectively) in the used filter media mentioned previously. The least amount of water-soluble Cu (13.1), Zn (1.6), Ni (7.71), As (23.6), and Pb (0.19) are extracted from MnR sediment. Comparison between the heavy metals indicates that water-soluble fraction of Pb is the lowest measured from all media investigated. The lowest water-soluble fraction of Pb is measured in the used media sampled from EBF laterite (0.3 µg/g) while 3 µg/g and 0.6 µg/g Pb are measured from KVA volcanic ash and DQ scoria respectively. The water-soluble fraction of Pb is mostly negligible in all filter media.

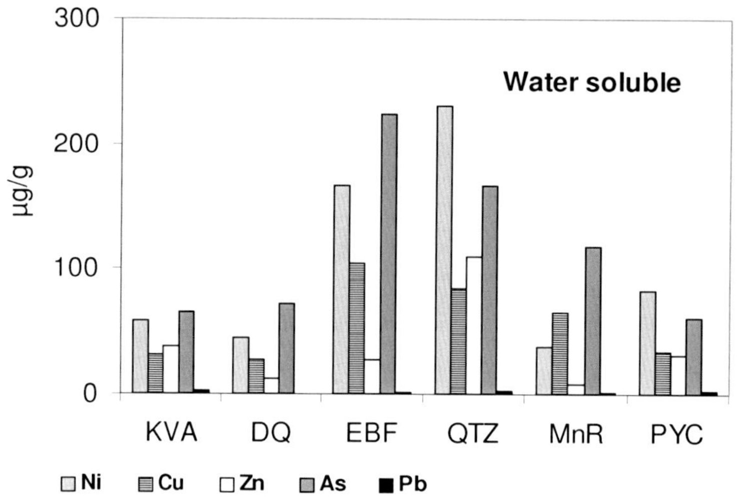

Fig. 5.23. Partitioning of heavy metals (0 - 10 cm depth section) in water-soluble fraction after water extraction procedure (modified DIN 38414 S4 1984).

Comparison of the results of single step water extraction procedure confirmed that the removal of most heavy metals from the water-soluble fraction is confined to the first 5 - 10 cm of the columns (Fig. 5.24). Between 5 and 10 cm depth section of the used KVA volcanic ash highest concentrations of Ni, Cu and Zn are observed. Distribution of the water-soluble fraction of Ni and As in the depth section of the used KVA volcanic ash, DQ scoria and EBF laterite show more or less a similar trend. In the used QTZ sand the distribution of Ni, Cu, Zn, As, and Pb is uniform throughout the depth profile while in KVA volcanic ash more than 60% of the heavy metals are bound in the first 5 - 10 cm of the column (Fig. 5.24). Based on the results of the single step water extraction scheme, release/mobility of Ni and As are higher than the other heavy metal studied.

Large amounts of Al, Fe, and Mn are desorbed from all filter media used after the leaching experiment. Concentrations up to 63 µg/g, 178 µg/g, and 40 µg/g of Al are measured in effluents of KVA volcanic ash, PYC pyroclastics, DQ scoria, and EBF laterite respectively. The Al concentrations desorbed are lower in MnR sediment (8.5 µg/g) while QTZ sand shows the lowest Al concentration (6 µg/g). The highest desorbed

Mn concentration (564 μg/g) is measured in MnR sediment, which is Mn oxide rich fluvial sediment. Desorbed concentrations of Mn are also elevated in the EBF laterite (176 μg/g) than that of the volcanic materials. Compared to Mn, Fe is released at lower concentrations from columns sludges of DQ scoria (12 μg/g) and EBF laterite (24 μg/g). The lowest desorbed Fe concentrations are measured in column sludges taken from QTZ sand (4 μg/g) and MnR sediment (5 μg/g) (Fig. 5.24d and e).

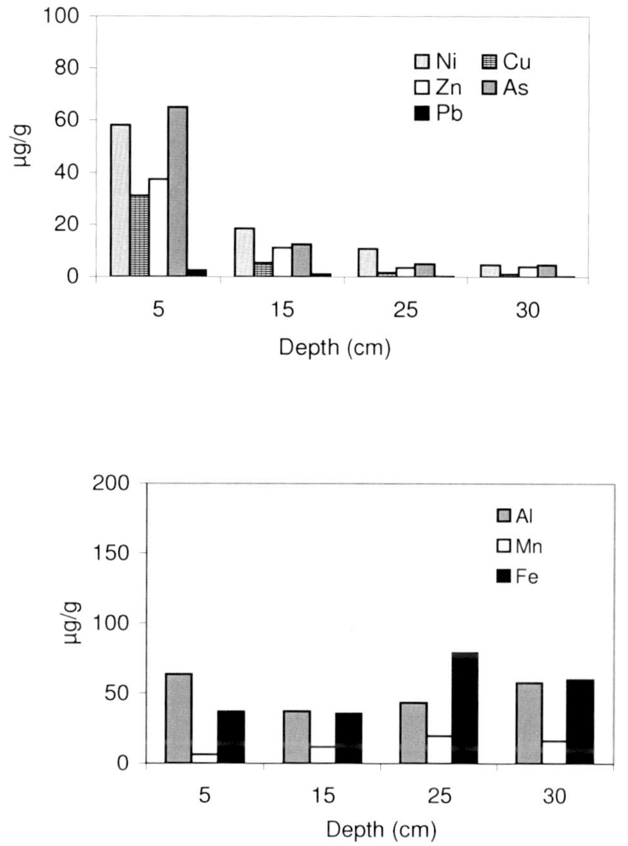

(a)

Fig. 5.24. Distribution of water soluble fractions of Ni, Cu, Zn, As, Pb, Al, Mn, and Fe in different depth sections (a) KVA volcanic ash, b) DQ scoria, (C) EBF laterite, (d) QTZ sand, (e) MnR sediment, and (f) PYC pyroclastics (After modified DIN 38414 S4 (1984)) leaching test).

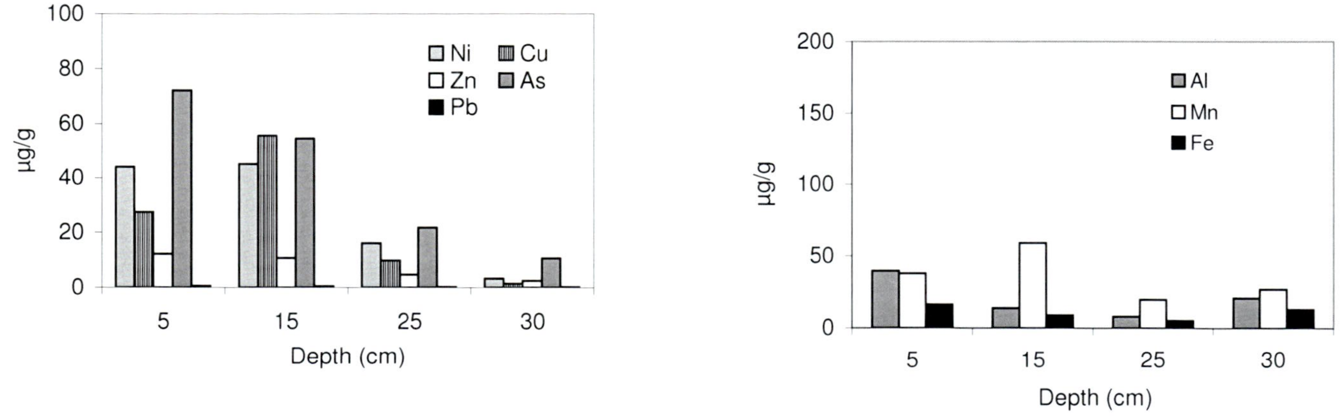

(b)

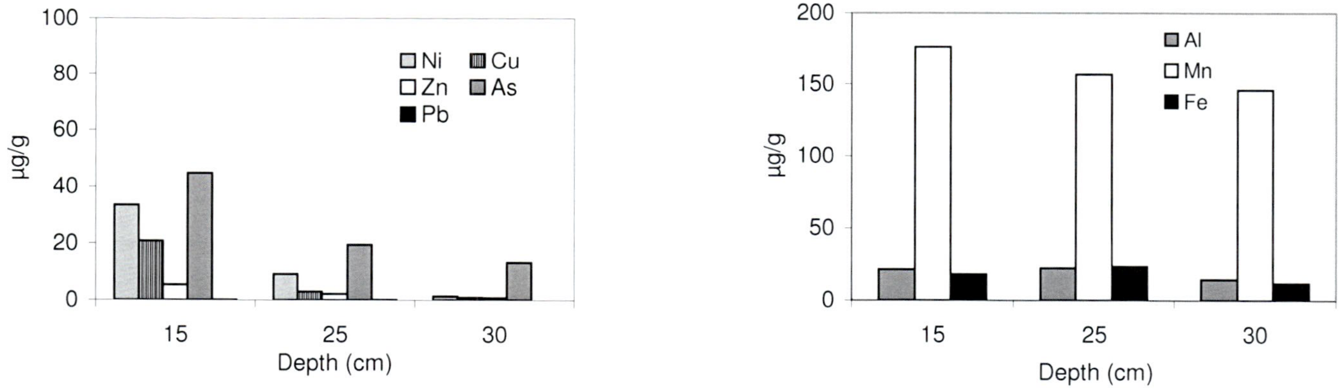

(c)

Fig. 5.24. Distribution of water soluble fractions of Ni, Cu, Zn, As, Pb, Al, Mn, and Fe in different depth sections *(contd.)*.

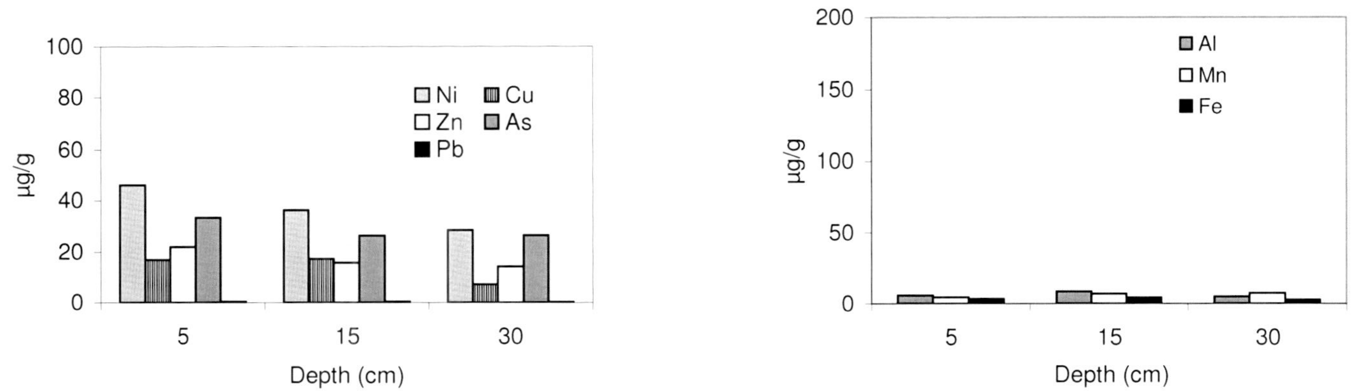

(d)

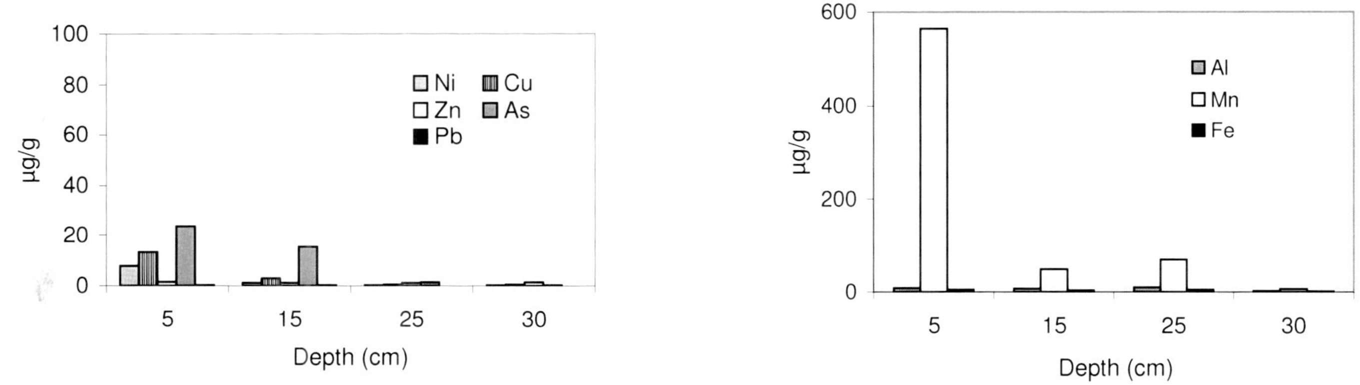

(e)

Fig. 5.24. Distribution of water soluble fractions of Ni, Cu, Zn, As, Pb, Al, Mn, and Fe in different depth sections *(contd.)*.

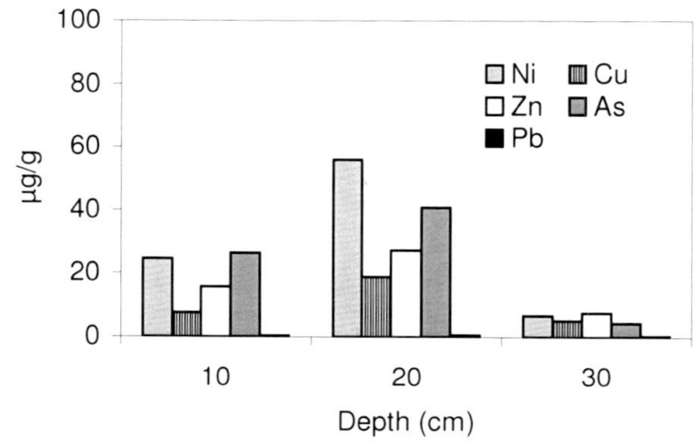

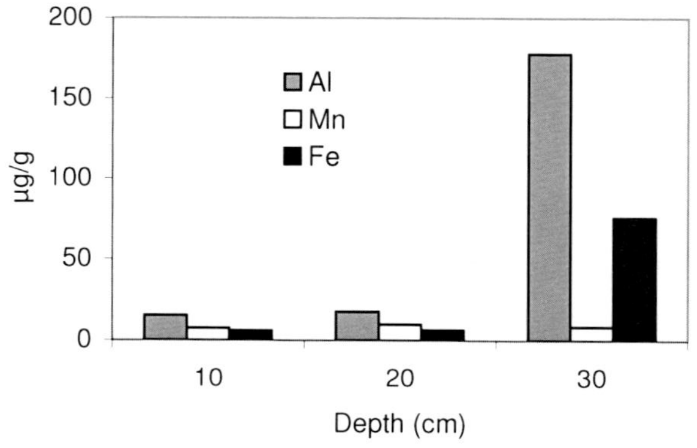

(f)

Fig. 5.24. Distribution of water soluble fractions of Ni, Cu, Zn, As, Pb, Al, Mn, and Fe in different depth sections (a) KVA volcanic ash, b) DQ scoria, (C) EBF laterite, (d) QTZ sand, (e) MnR sediment, and (f) PYC pyroclastics (After modified DIN 38414 S4 (1984)) leaching test).

The results of BCR sequential extraction of the filter media used are presented in table 5.8. The three BCR fractions of Cu, Ni, Zn, As, Pb, Fe, and Mn are compared in figure 5.25. In the sludge samples of KVA volcanic ash Ni, Cu, and Zn are mainly present in the exchangeable fraction (73%, 77% and 73% respectively) while 24%, 20% and 25% of these metals are present in the reducible fraction. The organically bound metals are less significant. Similarly sludge samples taken from DQ scoria and EBF laterite show higher association of Ni (80%), Cu (83%), and Zn (80%) in the exchangeable fraction while in PYC pyroclastics the exchangeable fractions are 56%, 73%, and 72% respectively. The exchangeable fractions of Ni, Cu, and Zn in column sludges of MnR sediment (80%, 70% and 78% respectively) are similar to that of DQ scoria and EBF laterite. In the sludge samples of all columns the reducible fractions of heavy metal show similar amounts (~ 20%) except for Pb. The reducible fraction of Pb accounts for 50 % in sludge samples of MnR sediment while KVA volcanic ash, DQ scoria, EBF laterite and PYC pyroclastics show 42%, 35%, 35% and 30% respectively. The oxidizable and water-soluble fractions of Ni, Cu, Zn, and As are below 5% for all column sludges except in PYC pyroclastics where 30% Ni and 13% Cu are present in the water-soluble fraction (Fig. 5.25 d, Table 5.8).

The fractionation of Pb in all column sludges is mainly attributed both to the acid exchangeable and reducible fractions. However, in sludges of PYC pyroclastics and QTZ sand a slightly higher fraction of Pb in the acid exchangeable fraction (64%) is measured. A somewhat different association of Pb in MnR sediment sludges is shown where Pb is almost uniformly bound to the reducible and oxidizable fraction. Compared to 5% of Pb of the acid exchangeable fraction, 53% of Pb is bound to the reducible fraction of MnR sediment sludges while the oxidizable fraction accounted for 42%. Unlike the other heavy metal, the association of Pb in the acid exchangeable fraction is negligible in sludges of MnR sediment (Fig. 5.25). Though, the fractionation of Pb in the oxidizable fraction is insignificant in sludges of KVA volcanic ash (6%), PYC pyroclastics (2.6%) and QTZ sand (3%), considerable amounts of Pb are bound in DQ scoria (14%) and EBF laterite (10%).

Like the other heavy metals, more than 80% of exchangeable As is released from all media except from MnR sediment. In column sludges of KVA volcanic ash and QTZ sand As is almost entirely present in the exchangeable fraction (90%). Only 56% of exchangeable As is released from MnR sediment while the reducible fraction accounts for 43%. Slightly lower exchangeable As is released from PYC pyroclastics (70%). The highest reducible fraction of As is measured in MnR sediment (43%) while in KVA volcanic ash, QTZ sand, and PYC pyroclastics, this fraction is below 5%. In EBF laterite and DQ scoria the reducible fraction of As reached 17% (Fig. 5.25). The oxidizable and water-soluble fractions of As in all column sludges are insignificant.

Table 5.8. Comparison between the BCR sequential extraction and water leaching tests in evaluating fractionation of heavy metals in the column sludge (µg/g).

Sample	Fraction	Ni	Cu	Zn	As	Pb	Mo	Cd	Al	Cr	Fe	Mn
KVA V ash	Acid exchangeable	1000	1700	2900	1050	580	4.20	6.07	380	115	2950	85.3
	Reducible	339	440	981	61.3	480	20.9	2.03	1928	5	2590	43.5
	Oxidizable	44.1	67.4	88.3	14.9	72.8	12.9	0.14	3270	39	1823	31.6
	Water soluble	58	31.12	37.42	65.08	2.58	1.14	0.22	63.34	0.13	37.02	6.42
DQ Scoria	Acid exchangeable	1800	1400	2500	900	300	3.86	4.20	6600	150	10900	970
	Reducible	291	233	396	184	216	10.1	1.08	10373	21	23649	444
	Oxidizable	89.6	51.3	61.4	13.5	83.7	4.04	0.02	9073	64	4398	158
	Water soluble	44.1	27.40	12.35	72.12	3.05	2.92	0.38	197.80	0.19	81.19	188
EBF laterite	Acid exchangeable	1100	1150	1420	860	350	3.01	1.95	1740	140	9500	2140
	Reducible	217	193	336	194	230	2.29	0.56	4624	10	15160	656
	Oxidizable	58.3	48.3	67.0	14.5	66.3	5.05	0.02	6591	50	2726	94.2
	Water soluble	33.4	20.78	5.35	44.88	0.25	0.46	0.06	21.5	0.05	18.16	176
Qtz sand	Acid exchangeable	800	1100	1300	670	350	3.58	2.11	500	160	3340	44.0
	Reducible	197	207	438	37.5	199	9.60	0.74	1228	4	1876	24.6
	Oxidizable	20.1	32.0	60.6	3.74	16.4	4.58	0.02	1392	45	585	9.01
	Water soluble	46.2	16.89	21.90	33.38	0.54	0.54	0.07	5.92	0.03	4.75	3.67
PYC V ash	Acid exchangeable	782	983	1295	772	338	3.95	1.77	928	151	3512	58.6
	Reducible	165	161	297	24.4	166	2.79	0.51	2815	4	1913	34.2
	Oxidizable	24.2	31.8	48.7	4.94	13.7	6.04	0.02	3063	48	759	22.1
	Water soluble	415	169	157	306	11.8	3.16	0.52	238	0.2	106	30.7
MnR sed.	Acid exchangeable	2443	2103	3834	2349	15.6	0.36	18.0	11127	115	1229	2443
	Reducible	556	779	913	1855	167	7.25	14.4	12500	32	30767	556
	Oxidizable	9.79	37.1	174	18.6	135	5.29	0.91	892	35	710	1314
	Water soluble	7.71	13.08	1.55	23.60	0.19	7.23	0.04	8.47	0.03	4.50	564

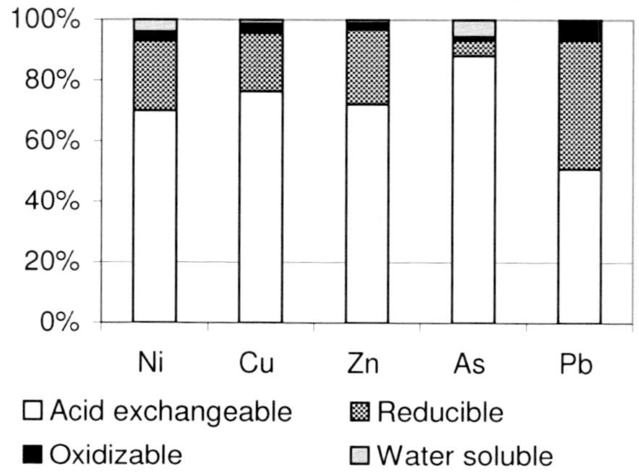

(a)

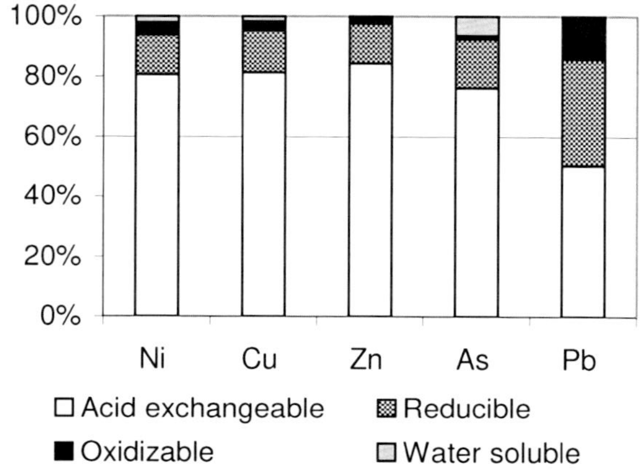

(b)

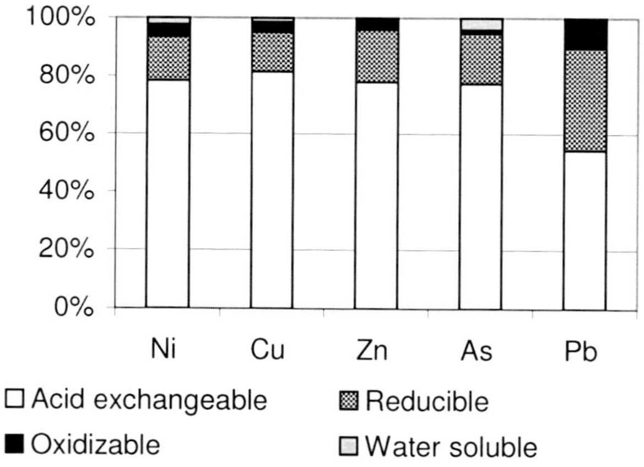

(c)

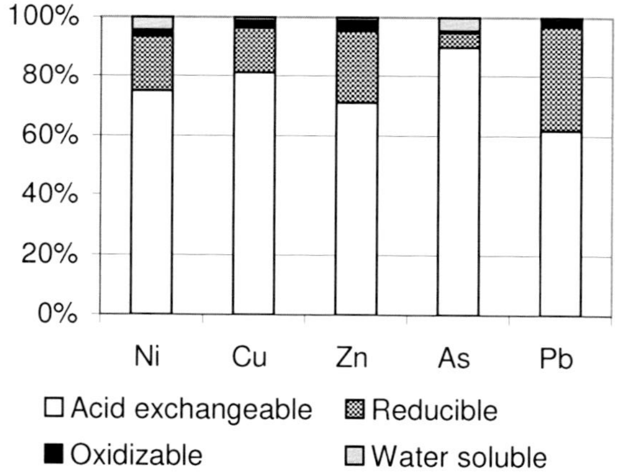

(d)

Fig. 5.25. Partitioning of heavy metals in water soluble, acid exchangeable, Fe/Mn oxide and organic bound fractions of (a) KVA volcanic ash, (b) DQ scoria, (c) EBF laterite, (d) QTZ sand, (e) PYC pyroclastics, and (f) MnR sediment (0 - 5 cm) as evaluated by BCR sequential extraction.

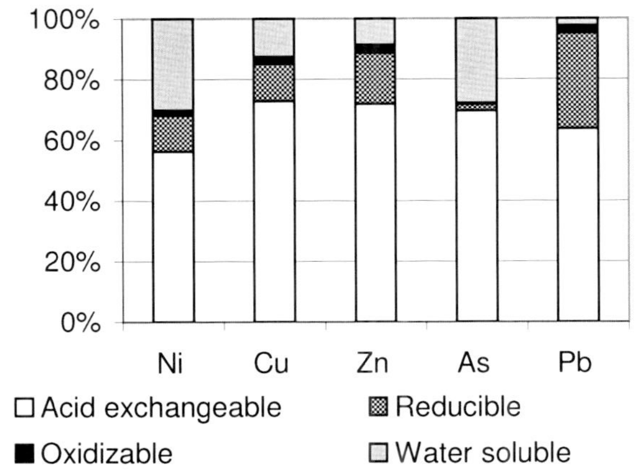

(e)

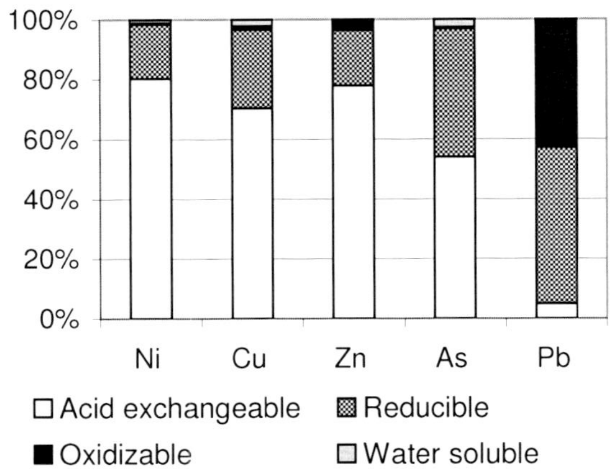

(f)

Fig. 5.25. Partitioning of heavy metals in water soluble, acid exchangeable, Fe/Mn oxide and organic bound fractions of (a) KVA volcanic ash, (b) DQ scoria, (c) EBF laterite, (d) QTZ sand, (e) PYC pyroclastics, and (f) MnR sediment (0 - 5 cm) as evaluated by BCR sequential extraction *(contd).*

Results of the BCR sequential extraction procedure also confirmed that larger proportion of heavy metals bound to the acid exchangeable fraction are confined to the first 5 - 10 cm depth of the columns. In sludge samples taken from KVA volcanic ash, MnR sediment and DQ scoria, almost all metals are retained within the first 5 - 10 cm of the column packs. The depth section distribution of Ni, Cu, Zn, As, and Pb are shown in figure 5.26. The highest desorbed concentrations of Ni (2443 µg/g), Cu (2100 µg/g), Zn (4000 µg/g), and As (2350 µg/g) are observed in sludge samples taken from MnR sediment (0 - 5 cm). Substantial amount of desorbed Ni (2070 µg/g), Cu (2500 µg/g), Zn (3820 µg/g) and As (1167 µg/g) are also measured from sludges taken from DQ scoria (0 - 5 cm). Unlike in the previously mentioned sludge samples, the depth section distribution of Ni, Cu, and Zn seemed to be even in sludge sample taken from PYC pyroclastics and QTZ sand (Fig. 5.26).

The major results of this work include (1) the water quality characterization of surface water and groundwater in the Akaki Sub-Basin. The type of water in the study area is alkaline earth metal fresh water with higher contents of alkali ions and prevailing hydrogen carbonate. Nitrate is measured in higher concentrations in groundwater samples than in surface water samples. 87% of the concentrations of NO_3^- in groundwater samples are two times higher than the US EPA MCL. Suspended particulate matter is the main carrier of heavy metals of concern such as Cu, Zn, Ni, As, and Pb. (2) The filter media removed all heavy metals of interest except As. Arsenic removal is not effective by all filter media investigated. Removal of As is the highest in MnR sediment (>90%) followed by DQ scoria (36%) while the other filter media are insignificant in removing As from polluted water. The main removal mechanisms involved are adsorptive filtration that involves the adsorption of soluble heavy metals on the surfaces of the filter media. Due to the biological layer (biofilm) formed on the surfaces of the filter media most heavy metals and other contaminants are removed on the first few centimeters of the columns media. As the simulated polluted water passed through the column media, suspended particulate matter and heavy metals bound to them will be removed by physical and chemical filtration processes.

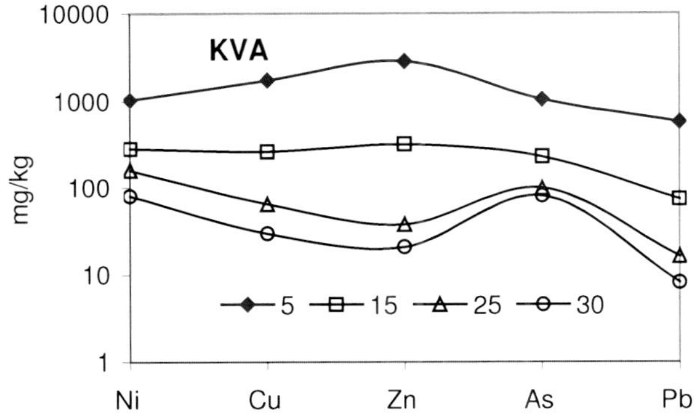

(a)

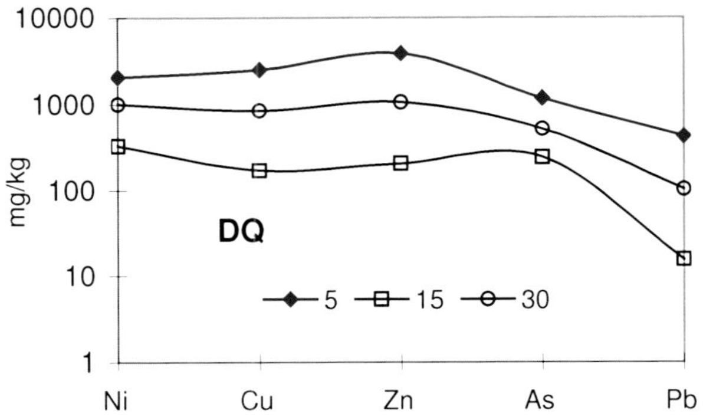

(b)

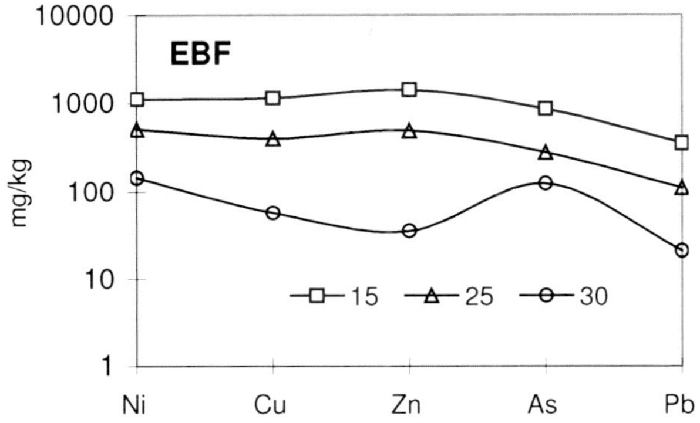

(c)

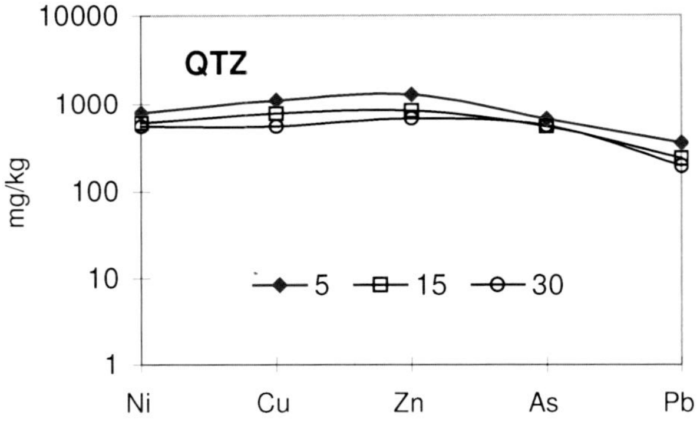

(d)

Fig. 5.26. Depth section distribution of exchangeable fractions of Ni, Cu, Zn, As, and Pb in (a) KVA volcanic ash, b) DQ scoria, (C) EBF laterite, (d) QTZ sand, (e) MnR sediment, and (f) PYC pyroclastics (after BCR sequential extraction).

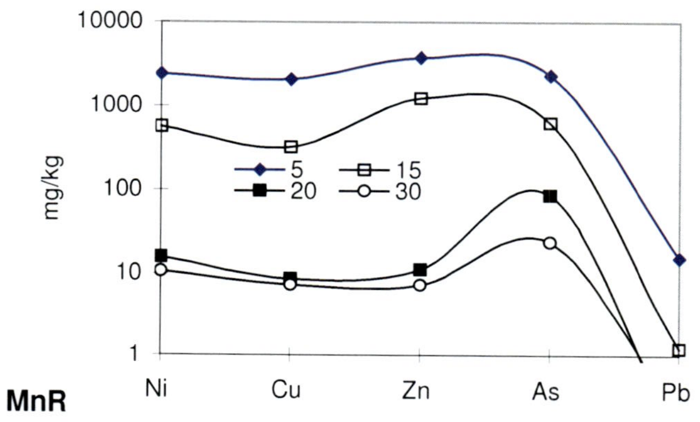

(e)

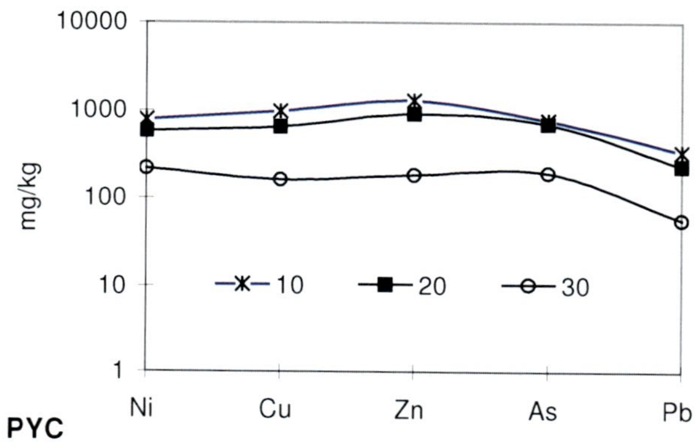

(f)

Fig. 5.26. Depth section distribution of exchangeable fractions of Ni, Cu, Zn, As, and Pb in (a) KVA volcanic ash, b) DQ scoria, (C) EBF laterite, (d) QTZ sand, (e) MnR sediment, and (f) PYC pyroclastics (after BCR sequential extraction).

6. Discussion

In this study, emphasis has been given to water quality characterization of the Akaki Sub-Basin in terms of inorganic nutrients and heavy metal contamination. Detail of the general physicochemical characterization of the surface and groundwater of the Akaki Sub-Basin are discussed. Some of the geochemical and anthropogenic factors that determine the overall chemical composition of the Akaki waters are discussed briefly. The most important contaminants of interest are Mn, Fe, Ni, Cu, Zn, As, Pb and Cd in dissolved as well as suspended particulate matter (SPM) bound form. In this regard, there are no previous research works that have been done in characterizing the distribution of these heavy metals and As on suspended particulate matters of the Akaki Sub-Basin water. In this work heavy metals removal capacities of local materials have shown promising outcome, which can be adapted to other areas where surface water and groundwater are polluted by heavy metals.

6.1 Water Quality Characterization

In the Akaki Sub-Basin, which is mainly characterized by volcanic terrain, water-rock interactions derived a hydrochemistry specific to the type of rock involved in the reaction. In headwater areas, such as Legedadi, Gefersa and the Entoto Ridges, ground and surface water tend to be relatively unpolluted showing less than 500 mg/l of dissolved solid (Alemayehu 2001). In the lower lying basin areas, such as Akaki, both surface and ground water tend to be polluted (more than 1,000 mg/l of dissolved solids) and contain Na^{+}, Ca^{2+}, Mg^{2+}, NO_3^{-}, HCO_3^{-}, and SO_4^{2-} as the major dissolved constituents. There are also water resources that have been directly or indirectly affected by anthropogenic pollution whereby the major chemistry is influenced mainly by the input of pollutants. The amounts of dissolved compounds or suspended solids in river water depend on the solubility of the rocks and soil and the rainfall within a drainage basin (Interlandi et al. 2003). The major ion composition of surface waters is controlled by the interaction of precipitation with surfacial geological and biological materials. In the Akaki Sub-Basin, the concentrations of chemical compounds are significantly affected by the weathering products of volcanic rocks as well as the circulation of thermal water, which is located at the center of Addis Ababa (Alemayehu 2001). The other sources are the anthropogenic inputs of domestic and industrial wastes that enrich certain anions and cations such as Na^+, Cl^-, NO_3^-, SO_4^{2-} and some heavy metals in surface and groundwater.

Specific electrical conductivity is one of the useful indicators of the total content of ions and of dissolved inorganic components (Hem 1985). Though most groundwater samples in the Akaki Sub-Basin show fairly normal values of electrical conductivity, some polluted groundwater samples such as Iv13 and Ah15 show elevated amounts (870 µS/cm) of conductivity. These amounts are above the Secondary Maximum Contaminants Levels (500 mg/l as TDS) set by the US EPA 1999 that indicates an increased concentration of major ions. The electrical conductivity upstream of the Small Akaki River near ELICO (El18) reach 1060 µS/cm that shows pollution from ELICO tannery in contrast to downstream part of the river at sample site Ju11 (516 µS/cm).

Most of the groundwater samples have water temperatures between 20 °C and 30 °C while some variation might be due to sampling duration. On clear sunny days, surface water samples are warmer than those collected on cool days. Except for Gerado pool Gr7 (T=29 °C) and the Akaki Textile effluent Ak14 (T=26 °C) all surface water samples show temperatures between 20°C and 22°C. The effluents of ELICO tannery (El19) also show high temperatures (T=27°C). Relatively higher water temperatures in the effluents are due to the release of wastewater from the two factories.

The Aba Samuel Lake shows concentration of DO as low as 2.4 mg/l. Dissolved oxygen as low as 0.2 mg/l is also measured in the Akaki River just before the junction to the Aba Samuel Lake. High Biological Oxygen Demand (BOD) and Chemical Oxygen Demand (COD), indicating the presence of organic contaminants (biodegradable and non-biodegradable) were found at the junction of the Aba Samuel Lake (13 mg/l) and the Small Akaki River (28 mg/l) (Alemayheu 2001). It is established that the origin of these high values are the high load of domestic and industrial wastewater discharged into the water systems. Low DO causes the water to change from oxidizing (aerobic) to reducing (anaerobic) conditions and also causes the oxidation-reduction potential to change from positive to negative values (Weiner 2000). Under this reducing condition, which is the case in the Aba Samuel Lake (Dk4), the Small Akaki River (Ju11, El18) and the Akaki textile wastewater (Ak14) less than 1 mg/l of NO_3^- concentrations are measured. In the Aba Samuel Lake, in spite of the fact that the concentration of dissolved oxygen (DO) is very low, variety of aquatic plants and animals exist.

Results of water parameter measurements indicated that the Aba Samuel Lake has undergone a high eutrophication process that has reduced the NO_3^- concentration (below detection limit, <1 mg/l). Plant productivity caused by nutrients is another factor of concern in the Aba Samuel Lake. In lakes, nutrient enrichment almost always increases algal production, a condition known as cultural eutrophication. One effect of eutrophication is that an abundance of algae may cause taste and odor problems in drinking-water supplies. A second effect of eutrophication is the increased uptake of dissolved oxygen by bacteria in response to higher concentrations of organic matter. If oxygen is taken up by decaying organic matter faster than it is imported from the atmosphere or produced by photosynthesis, it becomes depleted. Furthermore, oxygen depletion causes basic changes in the chemical environment that allow materials (including many metals) that are formerly precipitated or bound to sediments to become soluble and, therefore, mobile.

During the sampling campaigns, it has been also observed that the suspended load is much higher in the lake and the two river systems, which is resulted from surface runoff upstream of the study area. According to the US EPA (1999) standards for drinking water, domestic water supplies should have pH values in the range 6.5 to 8.5. Except in the El19 surface water sample (pH 6.8), the pH values for all surface and groundwater range between pH 7 and 8 ± 0.05 indicating that they are in accordance with the Secondary Maximum Contaminants Levels set by the US EPA (2000).

Alkalinity is the acid neutralizing capacity (ANC) of solutes in a water sample, reported in equivalents per liter (or milliequivalents or microequivalents per liter). Alkalinity consists of the sum of titratable carbonate and non-carbonate chemical species in a filtered water sample (filter membrane of 0.45 - µm pore size or smaller). Any substance in a water sample that reacts with strong acid titrant can contribute to the acid neutralizing capacity. Important noncarbonate contributors may include organic ligands (especially acetate and propionate) as well as hydroxide, silicate, borate, and less commonly ammonia and sulfide (Hem 1985). When found in unusually high concentrations, PO_4^{3-} and AsO_4^{3-} may also contribute to the acid neutralizing capacity of natural waters (Stumm and Morgan 1981). Except for unusual natural waters and waters substantially affected by human activity, noncarbonate ionized contributors generally are not present in large enough quantities to affect alkalinity or Acid Neutralizing Capacity. The principal source of CO_2 species that produce alkalinity in surface or ground water is the CO_2 gas fraction of the atmosphere, or the atmospheric gases present in the soil or in the unsaturated zone lying between the surface of the land and the water table (Hem 1985).

Alkalinity measurements on the Akaki water have shown that all samples lie in the alkalinity range of 2 - 2.5 mmoleq/l. Hydroxides and silicates being the common rock forming minerals in the Akaki area are the most important contributors of alkalinity. The calculated bicarbonate alkalinity is 250 mg/l. Naturally occurring levels of alkalinity reaching at least 400 mg/l as $CaCO_3$ are not considered a health hazard (US EPA 1992). EPA guidelines also recommend a minimum alkalinity level of 20 mg/l as $CaCO_3$. According to Weiner (2000), the natural background alkalinity is not reduced by more than 25% by any discharge. Table 5.1and 5.2 show how the water parameters vary among the different Akaki water samples.

Particulate matter can be an important contributor to alkalinity and must be removed by filtration before titrating the sample for the alkalinity determination. Since the particulate matter load of the Akaki surface water is higher, alkalinity measurements in unfiltered water samples can be much higher than what is measured in the filtered water samples.

Natural and human-influenced sources contribute chemicals to the dissolved solids of the Akaki River and its sub-basin. Based upon their electrical charge the chemicals are separated into two groups. The major positively charged chemical ions (cations) are Na^{+}, K^{+}, Ca^{2+}, and Mg^{2+}, and the major negatively charged chemical ions (anions) are NO_3^{-}, HCO_3^{-}, SO_4^{2} and Cl^{-}. The concentrations of the major ions are influenced by surface runoff. The general chemistry of runoff waters is best described using the reviews of world average river water composition that have been prepared over the past few years (Berner et al. 1987). The world average river concentration of major ions is approximately 100 mg l^{-1} (Berner et al. 1987), which is approximately 20 times greater than the concentration in rain. However, evaporation can increase the total dissolved solids in rainfall by a factor of two. Hence, the primary mechanism for elevating the ionic content of surface waters is mineral weathering.

Weathering of watershed is of primary importance in controlling surface water chemistry. The main source of major elements and ions (Ca, Mg, Na, K, HCO_3, SO_4, Cl and Si) in the Aba Samuel Lake is weathering of drainage basin lithology. Gibbs (1970) suggested that a simple plot of TDS versus the weight ratio of Na/(Na + Ca) could provide information on the relative importance of the major natural mechanisms controlling surface water chemistry: (1) atmospheric disposition, (2) rock weathering and (3) evaporation and fractional crystallization. Surface water samples with Na/Na+Ca ~ 0.41 and an average TDS (conductivity) ~ 746 mg/l are controlled predominantly by rock weathering.

Comparison of the concentrations of major cations and anions of the Akaki water in relation to that of the estimated mean composition of the World River waters (Hem 1985) and unpolluted upstream part of the Akaki catchement and the Akaki Sub-Basin are given in table 6.1. Concentrations of Na and K in surface waters of Akaki are ten times higher than those of the mean composition of the World Rivers. The highest concentration of Na in surface water is observed in industrial effluent at sample site Ak14 (70 mg/l) and Akaki River El18 (86 mg/l). Of all samples analyzed for their content of Na, surface water samples show the highest concentrations compared to the groundwater samples (Fig. 6.1). The concentration of K in all groundwater samples is also relatively low as compared to the surface water samples. In surface water samples the ratio between Na and K versus the total cations ($Na+K/TZ^{+}$) is 64% while groundwater samples show a ratio of 36% (Fig. 6.1). Anthropogenic as well as geogenic sources are the possible contributors to the high $Na+K/TZ^{+}$ ratio in surface water. The partial treatment of wastewater in Ak14 by applying some metal salts might be the possible source for the increased Na concentrations. On the other hand, domestic as well as industrial effluents released in to the Akaki Rivers increased Na concentrations above the background values. The highest concentration of K in Gr7 samples could be a result of water-rock interactions in the Gerado Pool. The underlying rocks are of volcanic origin. Since the volume of water in the pool is greatly affected by evapotranspiration, the concentration of K and other metals are higher. The enrichment of Na in groundwater samples such as De2, Be6, Tl12, Ls16, and Ak17 is due to the weathering of Na rich feldspars like albite ($NaAlSi_3O_8$). Since K containing feldspar is as resistant to chemical weathering compared to other feldspar types, the release of K due to rock-water interaction into groundwater may not be substantial. Thus, low concentrations of K are expected under normal geochemical conditions in groundwater. Elevated concentrations of K in groundwater samples are results of inputs from anthropogenic sources.

In surface water, Mg and Ca concentrations are also two to three times higher than the mean composition of world river waters (Table 6.1). Magnesium ions are also contributed by geogenic as well as anthropogenic sources. The ratio of Ca and Mg versus the total cations ($Ca+Mg/TZ^{+)}$ is 37% in surface water samples while groundwater samples show 63% (Fig. 6.1). In igneous rock, Mg is typically a major constituent of the dark-colored ferromagnesian minerals. Specifically, these include olivine, pyroxenes, amphiboles, and the dark-colored micas, along with various less common species. Rock-water interactions of volcanic rocks covering wide areas of the Akaki Basin are responsible for the high Mg concentrations in groundwater compared to surface water.

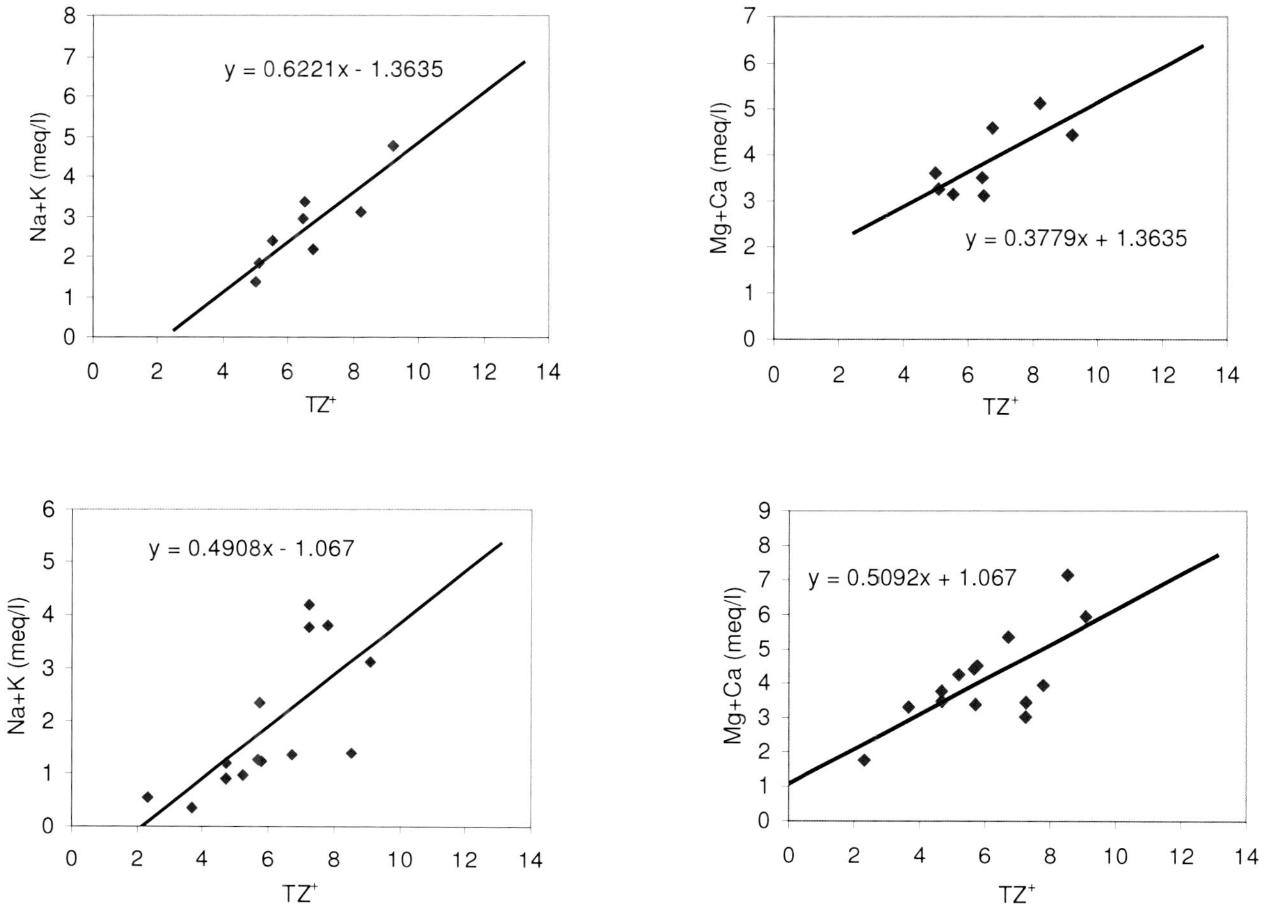

(a) (b)

Fig. 6.1. The ratio between Na+K vs. total cation (TZ^+) and Mg+Ca vs. total cation (TZ^+) (a) surface water, (b) groundwater.

Since the Akaki surface and groundwater are dominantly characterized by substantial Ca and Mg contents, the major cationic compositions comply well with those expected in many igneous terrains where weathering of igneous minerals, especially of the chain silicates pyroxene, amphibole, and the feldspars are common (Taylor and McLennan 1985). The plagioclase feldspar group of minerals represents mixtures in various proportions of the end members albite, $NaAlSi_3O_8$, and anorthite, $CaAl_2Si_2O_8$. Some Ca^{2+} is, therefore, expected in water that has been in contact with igneous and metamorphic rocks. Carbonic acid weathering of silicate rocks is also an important source of Ca^{2+}, HCO_3^-, K^+, Mg^{2+}, Na^+ and dissolved Si.

The (Ca+Mg)/(Na+K) molar abundance ratio in silicates of the upper crust is generally 1.0 (Taylor and McLennan 1985). This ratio in the Aba Samuel Lake is 1.2 indicating that the cations are derived from silicate weathering. For the Akaki River at sample site Ju11 the ratio is 2.1 implying that the sources of the cations are not only silicate weathering. However, this ratio is more than 2.0 in 64% of the groundwater samples indicating that other than water rock interactions, non-natural sources also contribute to the higher contents of the cations.

The sources of NO_3^-, Cl^- and SO_4^{2-} in surface water and groundwater can be broadly divided as natural and anthropogenic. Natural sources are generally ubiquitous; however, their contribution is usually low because, over the course of time, natural systems have established balances between the production and consumption of nutrients. Anthropogenic sources arise from many activities.

Nitrate ions generally are derived from the oxidation of organic matter with a high content of protein. Their presence may indicate a pollution source and their occurrence is usually associated with shallow groundwater sources. Nitrate concentrations in fresh water do not generally exceed 5 mg/l, although in rural areas, where NO_3^- fertilizer is liberally applied, concentrations may exceed 600 mg/l (Bell 1998).

Table 6.1. Comparison of estimated mean composition of river water of the world with selected surface water samples in the Akaki Sub-Basin (modified after Hem 1985).

Dissolved constituent	World rivers	Yeka Spring[1] Up-stream Akaki	Dk4	Ku1	Dk5	Gr7
	mg/l	mg/l	mg/l	mg/l	mg/l	mg/l
Ca	15	6	40	37.8	40	27.5
Mg	4.1	21	16	15	14	23
Na	6.3	8.2	50	62	46	14
K	2.3	-	29	26	15	47
SO_4	11	-	40.1	17.3	28.2	7.4
Cl	7.8	2.2	54.6	31	45.6	10.9
NO_3	1	55.3	0.0	1.84	1.02	0.0
PO_4	0.08	0.11	-	0.13	0.06	0.04

[1] Relatively unpolluted upstream part of the Akaki catchement

Concentrations of NO_3^- in the surface waters of Akaki are ten times higher than those of the mean composition of the World Rivers in 40% of the total samples (n=8). In the Akaki Sub-Basin, drinking-water standards are exceeded for NO_3^- in both surface and groundwater. The Secondary Maximum Contaminant Limit (SMCL) for NO_3^- (10 mg/l; U.S. EPA 2000) is exceeded in 9 out of 14 groundwater samples (64%). In 3 of 8 surface water samples (37.5%) NO_3^- exceeded the SMCL. Ground water containing 3 mg/l or more NO_3^- is considered to be affected by human activity (Madison and Brunett 1984).

Except in the Akaki well field, most groundwater wells are shallower than 50 - 60m in depth. The wells are directly affected by NO_3^- pollution from the nearby surface water bodies that are loaded with anthropogenic pollutants. Tributaries upstream of the Small Akaki River have NO_3^- concentrations up to 700 mg/l (Lideta Spring), 500 mg/l (Ras Mekonnen Spring), and 200 mg/l (Addisu Gebeya). In most cases, the tributaries of the Small Akaki River are affected by direct disposal of domestic sewage and industrial effluents (Alemayehu 2001), which resulted in elevated concentrations of NO_3^-. In the upper and middle courses of the Akaki River drainage, farmers increase the productivity and yield of their farms by use of animal dung and chemical fertilizers. If more animal dung fertilizers are applied than are used by the crops, they can be transported into surface water and groundwater and become a major source of nutrients. The other main source of NO_3^- in the Akaki surface water and groundwater is human sewage where no wastewater treatment is applied. In addition to the nutrients derived from human sewage, municipal wastewater also contains nutrients from such sources as household cleaners and detergents.

Chloride concentrations are well below the US EPA SMCL (250 mg/l) in both surface water and groundwater samples. However, concentrations of Cl^- in the surface waters of Akaki are five times higher than those of the mean composition of the World Rivers in all of the samples (n=8). The groundwater Cl^- concentration is less than that of the surface water by approximately 10 mg/l at the Dk3 sample site.

The highest Cl^- concentration is detected in the Aba Samuel Lake (Dk4) (54 mg/l) while the groundwater at sample site Dk3 shows the maximum concentration of Cl^- (40 mg/l). Concentrations of Cl^- close to 50 mg/l are also detected in the outlet of the Aba Samuel Lake (Dk5) and the Small Akaki River at sample sites Ju11 and El18. The effluent samples taken from ELICO tannery also show concentrations close to 50 mg/l.

The Cl^- content of most groundwater may be due to the presence of soluble Cl^- from rocks and/or contamination by industrial effluent or domestic sewage. Though Cl^- is a minor constituent in the earth's crust, igneous rocks contain sodalite as an essential constituent that in dissolution contributes to Cl^- ions. Usually, the concentration of Cl^- in groundwater is 30 mg/l (Bell 1998). Elevated concentrations in the Akaki Sub-Basin may be the result of anthropogenic pollution.

Sulphate in the Akaki Sub-Basin may have geogenic and anthropogenic sources. High concentrations could be caused by release from agricultural runoff or seepage. In surface water the highest SO_4^{2-} concentrations are detected in the Small Akaki River at ELICO effluent El18 and El19 (43 mg/l) and Aba Samuel Lake at site Dk4 (40 mg/l). In 90% of the total surface waters samples of Akaki (n=8) concentrations of SO_4^{2-} are three times higher than those of the mean composition of the World Rivers. The sampling sites represent areas where domestic and industrial waste effluents are discharged indicating anthropogenic sources of SO_4^{2-}. The contribution of S to rivers from anthropogenic sources is assumed to reach about one-half as much as from natural sources (Berner 1971). The concentration of SO_4^{2-} ions in water can be affected by SO_4^{2-} reducing bacteria, the products of which are H_2S and CO_2. Hence, a decline in SO_4^{2-} ions is frequently associated with an increase in HCO_3. The concentration of SO_4^{2-} in groundwater usually is less than 100 mg/l, and may be less than 1 mg/l if SO_4^{2-} reducing bacteria are active (Bell 1998). In groundwater of Akaki the highest SO_4^{2-} concentrations are measured at sample sites of De2 and Dk3 indicating the pollution effect of the Aba Samuel Lake and the Akaki River.

The chemical characteristic of the Akaki River is influenced by the chemical composition of its tributaries that influence the river water composition. There are some perennial streams in Addis Ababa that are part of the catchement: Little Akaki, Bantyiktu, Kurtume, Kebena, and Big Akaki. The major anion and cation contents of these streams indicated that the streams have been affected by anthropogenic pollution. Alemayehu (2001) reported concentrations of Na^+ (147 mg/l), K^+ (15 mg/l), Ca^{2+} (48 mg/l), and Mg^{2+} (18 mg/l) in the Small Akaki River at sample site Behere Tsighe that corresponds well with our data. In this river, anthropogenic pollution indicators such as PO_4^{2-} concentrations are about 3 mg/l. In the Bantiyketu stream, the concentrations of Na^+, K^+, Ca^{2+} and Mg^{2+} are 30 mg/l, 4 mg/l, 10 mg/l, and 5 mg/l respectively while Cl^-, SO_4^{2-}, and PO_4^{2-} concentrations were 50 mg/l, 14 mg/l and 2.2 mg/l respectively. The Kurtume stream has relatively low major cations concentrations (Na^+ 30 mg/l, K^+ 4 Mg/l, Ca^{2+} 10 mg/l, and Mg^{2+} 5 mg/l) as compared to the other tributaries. Downstream from the Kebena stream (Urael) concentrations of Na^+, K^+, Ca^{2+} and Mg^{2+} reach 28 mg/l, 17 mg/l, 36 mg/l, and 15 mg/l respectively.

The chemistry of the Akaki River water can be characterized by a Ca - Mg - Cl and Ca - Mg - HCO_3 type. However, a hydrochemical shift of the water chemistry towards the SO_4^{2-} and Cl^- types is mostly related to the influence of contaminants into the subsurface rather than the natural dissolution processes. This is due to the fact that most of the springs located in the upstream parts of the Akaki catchment (less vulnerable to pollution) are characterized by high HCO_3^- concentration rather than Cl^- and SO_4^{2-}.

The relatively higher concentrations of HCO_3^- and CO_3^{2-} in the surface waters of the Akaki Sub-Basin are attributed to the anthropogenic pollution contributed by household and industrial effluents. Most of the surface water belongs to Na - HCO_3 and Na - Ca - HCO_3 types. The relative abundance of the ions can be related to geochemical reactions that have taken place as the water comes in contact with different minerals of the prevailing rock types. Moreover, there is also a significant contribution of elements due to pollution of the surface water and groundwater. As a result, some scattered data points are observed in the piper diagram (Fig. 5.4). Therefore, the surface and groundwater chemical composition represent the influence of natural as well as anthropogenic-induced materials.

Chemical analysis of surface water and groundwater samples collected from the Akaki Sub-Basin reveal that Na^+ and Ca^{2+} are the dominant cations. The major sources of Ca^{2+} and Na^+ cations in unpolluted surface water and groundwater are the minerals that constitute basic and acidic volcanic rocks. Rocks containing Na and Ca are more or less susceptible and upon weathering yield the metal cation and in solution. The resulting water from mafic volcanic rocks tends to be a Ca - Mg - HCO_3 while the dissolution of acidic volcanic rocks produces Na - HCO_3 water with relatively large amounts of SiO_2.

From the piper diagram (Fig. 5.4) it is also evident that the HCO_3 ions are the dominant anions in groundwater samples with variation in composition of Ca and Mg. The resulting groundwater types also vary from Ca/Mg - HCO_3 to Na - HCO_3 depending on the rock-water interaction that prevails in the study area. The relative abundance of these cations (Na and Ca) indicates the dissolution of minerals that constitute either mafic or acidic volcanic rocks.

In most cases, surface water is hydraulically connected to groundwater, but the interactions are difficult to observe, to measure and have been ignored in water management considerations and polices. Surface water interacts with groundwater in all types of landscapes as loosing or gaining streams or both. The environment where the polluted Akaki River interacts with the underlying groundwater varies from place to place. The water level in the Akaki River is located at higher elevation than the water level in the aquifer and hence is acting as a discharging river by releasing water in to the near by alluvial and volcanic aquifers. Elevated concentrations of total dissolved solids and NO_3^- are some of the indicators that the river - groundwater interaction, affects the shallow groundwaters close to the Akaki River.

6.2 Assessment of Heavy Metal Contamination in Akaki Water

It has been widely recognized that the fate and transport of metals within aquatic system is determined by the nature of their association with suspended particulate matter. The fate and transport of heavy metals and inorganic anions in the Aba Samuel Lake and the surrounding water systems are governed by the chemical composition of all the streams that flow into the lake and the geology of the catchment. Certain rock types contain elements that may be harmful to humans or animals if they are released and concentrated in water. In the Akaki Sub-Basin, heavy metals, As and U are detected in some water samples at concentrations that exceeded drinking-water standards.

Particulate matter is derived primarily from rock weathering processes, both physical and chemical, and may be further modified by soil-forming processes. Erosion subsequently transfers sediments or soil particles from their point of origin into freshwater systems (Chapman 1996). Two major natural sources of sediment to rivers and lakes can be considered: (i) products of continental rock and soil erosion, and (ii) the autochthonous material, which is formed within the water body and usually results from the production of algae and the precipitation of a few minerals, mostly calcite (Campy and Meybeck 1995).

In regions of very high mechanical erosion, the elemental content of river suspended matter reflects the principal origins. Depending on the lithological nature of the parent rock, the composition of river SPM is generally close to the composition of the parent rocks and it may present some variations in major elements (Table 6.2). In table 6.2 concentrations of Ca, K, Fe and Mn bound to SPM are compared between the Akaki River and tropical and arid zone basins. The same comparison has also been made between the Akaki River and basalt river basin (Chapman 1996).

Table 6.2. Comparison of Fe, Mn, K, and Ca bound to SPM between Akaki surface water and natural river water.

Surface water	Fe	Mn	K	Ca
Ak14	1433	53.7	32.3	330
Dk5	16000	837	358	5013
Dk4	5065	12020	7325	74205
El18	21.84	0.91	6.23	108
Ku1	74213	1200	539	8573
Gr7	6187	221	115	534
Ju11	8361	137	60	961
Basalt river basin water[1]	52600	1300	35400	19300
Tropical and arid zone basins water[1]	61700	890	7500	18300

[1] *Martin and Meybeck, 1979; Meybeck, 1988*

Compared to waters of tropical and arid zone basins higher Mn concentrations bound to SPM are found in the Aba Samuel Lake (Dk4) and the Akaki River (Ku1) (Table 6.2). The Akaki River at Ku1 has higher concentration of Fe bound to SPM than the basaltic river basin, tropical and arid zone basin waters. Potassium and Ca bound to suspended particulate matters are much higher in the Aba Samuel Lake (Dk4) as compared to the tropical and arid zone basin waters (Table 6.2). The Akaki River particularly is enriched with all the heavy metals indicated in the table than the corresponding tropical and arid zone basin waters described by Martin et al. (1979) and Meybeck (1988). According to Martin and Meybeck (1979) and Meybeck (1988) enrichment factor (EF) is given for each element. Here, the EF defined as the ratio of the surface water concentration to the average abundance for the tropical and arid zone basins water for each element. The enrichment factors calculated for Fe and Mn are 1.2 and 1.3 respectively indicating that Fe and Mn have been introduced into the Akaki River (Ku1) from anthropogenic sources. Except for Fe, the Aba Samuel Lake also shows elevated amounts of Mn, K, and Ca that have enrichment factors of 14, 1 and 4 respectively. The higher enrichment factors are indication that the lake has been enriched with major elements contributed not only from geogenic sources but also from anthropogenic sources. Upstream of the study area the Small Akaki River has the lowest concentrations of all major elements described before (Table 6.2 and 6.3). All the other surface water show lower enrichment factors in comparison to the Aba Samuel Lake and the Akaki River.

Table 6.3. Enrichment factors Fe, Mn, K, and Ca in Akaki surface water compared to tropical and arid zone basins water calculated after Martin and Meybeck (1979), Meybeck (1988).

Surface water	Fe	Mn	K	Ca
AKt14	0.02	0.06	0.004	0.02
Dk5	0.26	0.94	0.05	0.27
Dk4	0.08	13.5	0.98	4.06
El18	0.00	0.001	0.001	0.006
Ku1	1.20	1.35	0.07	0.47
Gr7	0.1	0.23	0.02	0.03
Ju11	0.14	0.15	0.01	0.05

When chemical alteration exceeds mechanical erosion, the most soluble elements are carried in the dissolved phase as ions (Ca^{2+}, Mg^{2+}, Na^{+} K^{+}) and dissolved SiO_2, whereas the least soluble ones (Al, Fe, Ti, Mn) remain in the soil that gradually becomes more enriched.

The composition and concentration of particulate matter in the aquatic environment is affected by the source and pathway of sediment input (Eisma 1993, Webster et al. 1990). Based on the geology, weathering, and hydrogeochemical conditions prevailing in the Akaki Sub-Basin the composition of the suspended particulate matter is more dominated by clay of montmorillonite type, oxides of Fe/Mn and organic matter.

The importance of suspended particulate matter (SPM) in pollutant transport in rivers in particular, its role in metal speciation in natural waters is well known and widely reported (Muller et al. 1990, Lynn 1999). In rivers, the behavior of SPM depends on particle dimensions. In aquatic systems two main fractions of SPM can be distinguished (i) the particulate fraction (diameter >1 µm) which is quite quickly settling and transporting metals to the sediments, and (ii) the colloidal fraction (diameter <1 µm) which remains suspended longer and may facilitate the transport of adsorbed species over appreciable distances. The colloidal fraction of SPM, usually defined as particles <1 µm in diameter (IUPAC), is the major transporting agent for heavy metals because of its large specific surface area and hence high potential for the adsorption of trace metal ions (Gustafsson et al. 1997). The highest concentrations of SPM up to 7 mg/l are measured in the groundwater samples of Akaki at sample site Be8. On the other hand, the low lying parts of the Akaki Sub-Basin show the highest surface water concentrations of SPM compared to the upstream parts. Compared to the up stream parts of the basin, the release of industrial effluents have been much higher in the low lying parts, which contributed to the high load of suspended solids and other contaminants. While the concentrations of SPM in surface water samples are not varying in wide ranges in groundwater samples variations up to two orders of magnitude has been observed between samples (Fig. 5.10). These variations could be due to aquifer lithological differences between groundwater sites, water-rock interactions, well depth and other factors.

Certain environmental factors influence the degree in which heavy metals and As are associated with SPM. The concentration and stability of SPM/colloidal matter in aquatic system are determined by the simultaneous effects of factors such as SPM concentration, (Honeyman and Santschi 1989, Filella and Buffle 1993, Gustafsson and Gschwend 1997) salinity, pH, oxygen saturation. Changes in pH can potentially affect SPM / colloidal stability via an effect on the surface charges of SPM and colloids. The stability of Mn-Fe oxides (main component of SPM) is controlled by the redox condition in aquatic system. Under suboxic to anoxic conditions, the insoluble Mn (IV) and Fe (III) species can be reduced to the soluble Mn (II) and Fe (II) forms.

Once released heavy metals would be associated with high concentrations of suspended particulate matters and/or colloids in rivers and lake systems. Particulates

and dissolved organic matter and colloids play very important roles in water chemistry and the binding and transport of metals and organic chemicals. One of the controlling mechanisms are sorption interactions that are discussed in Stumm and Morgan (1981), Whitfield and Turner (1987), Newman and McIntosh (1991), and Baker (1991a, b).
SPM containing high concentrations of organic matter and certain clay minerals (e.g. montmorillonite with a high cation exchange capacity) has the ability to sorb significant quantities of chemical components (Louma 1989). However, because of the large surface areas of silts, clays, and organic matter and the scavenging nature of oxides, SPM is also the major transport medium for heavy metals in rivers (Louma 1989, Sinclair et al. 1989). Particulate size and resulting total surface area available for adsorption are both important factors in adsorption processes and can affect metal bioavailability (Luoma 1983, 1989). Small particles with large surface-area-to-mass ratios allow more adsorption than an equivalent mass of large particles with small surface-area-to-mass ratios.

In the Akaki water, most heavy metals of interest are associated with the suspended particulate matter. The concentrations of Cu, Zn, Ni, As and Pb bound to SPM in groundwater samples De2, Dk3, Be8, E10 are higher than that of the surface water. This might be explained by the dominant proportion of the smaller size fractions of SPM in groundwater than in surface water. Clays and organic floccules (common constituents of fluvial SPM) tend to concentrate in the smaller size fractions (Horowitz et al. 1990) and adsorb relatively more contaminants per unit volume or mass due to their relatively larger surface-to-mass ratios. For instance, in slightly basic and anoxic marsh sediment environments, Zn is effectively immobilized and not bioavailable. The pH condition (pH = 7 - 8) persists in most natural water favors the low solubility of Zn and the other heavy metals regardless of the redox conditions (Gambrell et al. 1991). In surface water and groundwater of Akaki, the existing pH condition also favors for the general low solubility of most heavy metals.

The concentrations of Cu bound to SPM in E10, Ah15, and Ab22 water wells are the highest measured 465 µg/g, 18 µg/g 7, and 155 µg/g respectively. All these water wells are used particularly for drinking purposes. Since the water is used without any further purification by large rural communities, the contaminants on the suspended particulate matter and dissolved phases will get in to the food chain easily. Therefore, the presence of the investigated heavy metals in such significant amounts in the suspension phase contributes to one of the several causes of health risks in the Akaki area.

High Cu, Zn, Ni, As and Pb concentrations in the SPM can result from adsorption on to oxides of Fe, Mn clay minerals and organic matter (Benoit et al. 1994). Comparison of heavy metals bound to SPM with those of Fe, Mn, Ca, and K are made in figure 6.2 and 6.3 for surface water and groundwater respectively. During the sampling period of April 2003, almost all of the Cu, Ni, As and Pb loads were associated with SPM in all of groundwater and some of the surface water samples. Higher association of Ni, Cu, Zn, As and Pb is shown for Mn bound to SPM that has concentrations ranging between 10^2 µg/g and 10^4 µg/g compared to concentration of Fe bound to SPM (10 - 10^6 µg/g). This is an indication that in most waters of the Akaki Sub-Basin Cu, Zn, Ni, As and Pb bound to SPM are higher in the Mn oxide fraction of SPM than in the Fe oxide fraction though the concentrations of Fe bound to SPM are much higher than that of Mn. On the other hand, there is an increasing trend of Zn, Cu, Ni, As and Pb bound to SPM with increasing concentrations of Mn bound to SPM (Fig. 6.2). Though there is no clear indication of an increasing trend of these heavy metals and As in relation to Ca and K bound to SPM, higher association of heavy metals and As with Ca and K bound to SPM is also shown.

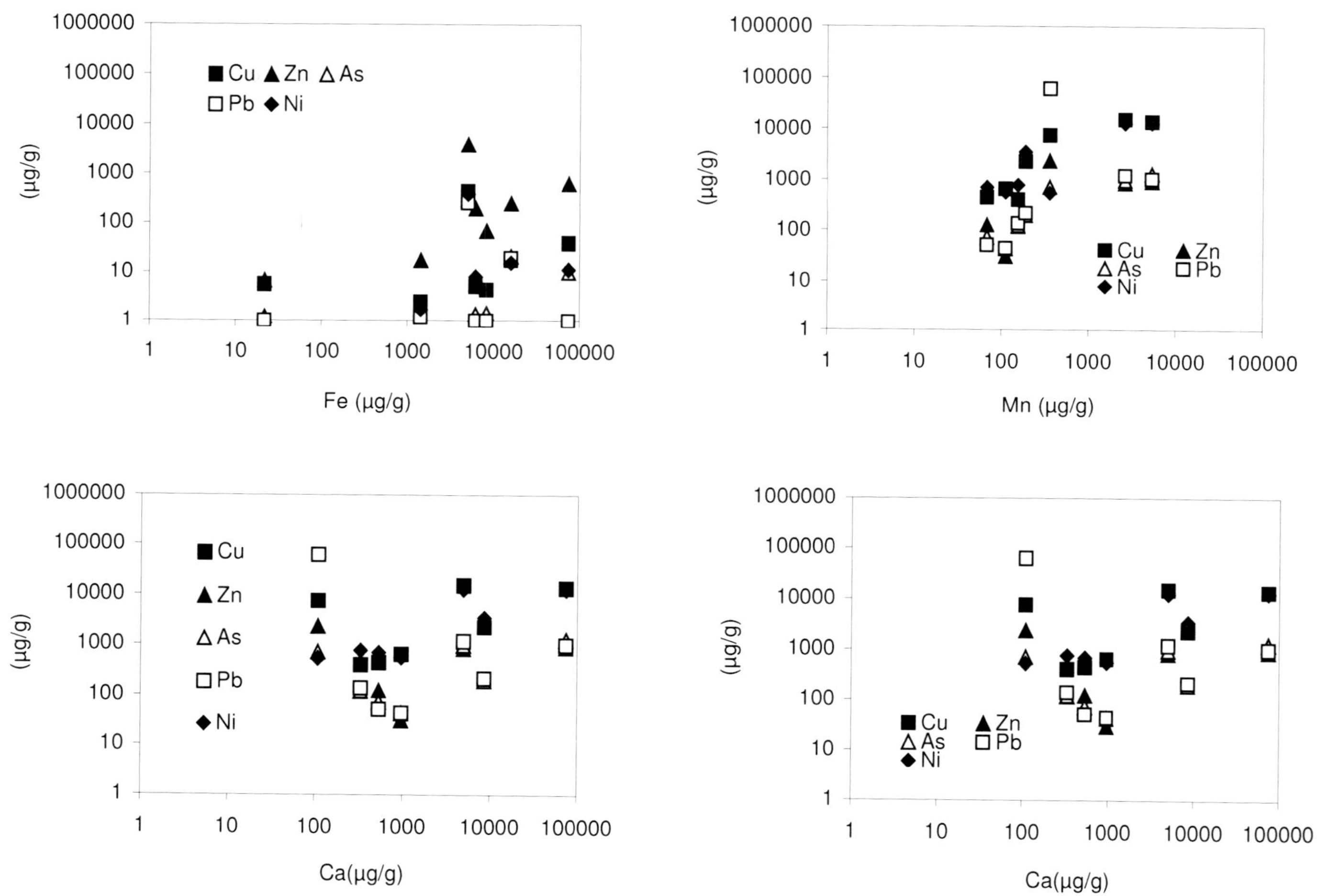

Fig. 6.2. Heavy metals bound to SPM vs. (a) Fe, (b) Mn, (c) Ca, and (d) K bound to SPM in Akaki waters (Surface water).

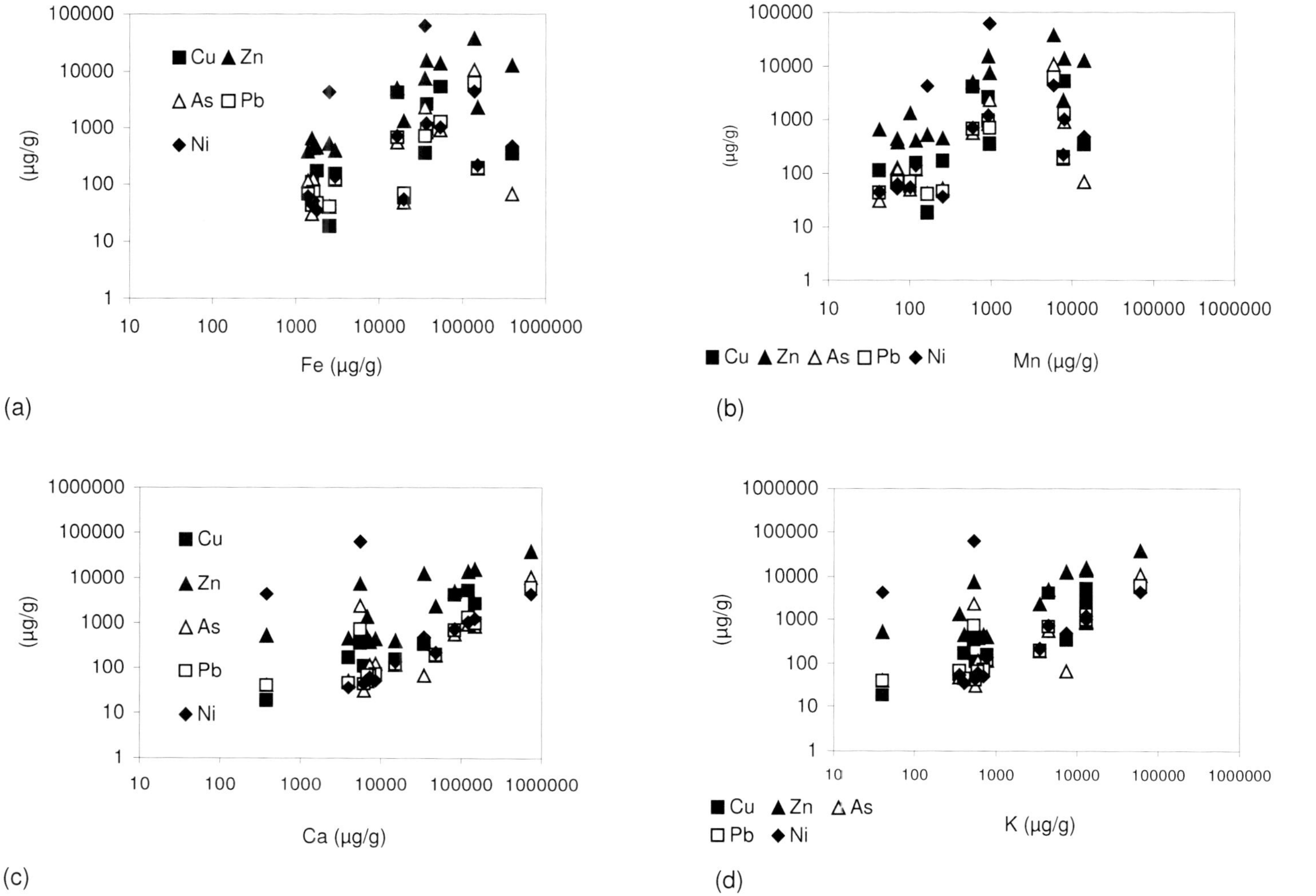

Fig. 6.3. Heavy metals bound to SPM vs. (a) Fe, (b) Mn, (c) Ca, and (d) K bound to SPM in Akaki waters (Groundwater).

The strong association of Zn, Ni, As, Pb and Cu with Fe, Mn, Ca, and K bound to SPM in groundwater samples reveals a positive relationship indicating that Fe oxides, Mn oxides, silicates, and clay minerals could be the most important components of SPM that bound the heavy metals in the Akaki waters. This is also confirmed with our SEM data, which reveal that Si, Ca, K, Fe, Mn and organic matter are the main elements detected in the SPM. In aqueous systems, there are four major solid phases that can be considered most relevant for metal sequestration: (i) carbonates, (ii) oxyhydroxide minerals (Fe and Mn), (iii) organic matter (live and dead) and (iv) sulfide minerals (Tessier et al. 1979). Among these solid phases, the most important ones that are found in the Akaki waters are oxides/oxyhydroxides of Fe and Mn and organic matter.

The Fe and Mn oxyhydroxides [e.g. Fe(III) oxides, FeOOH(s) and Mn(III,IV) oxides, $MnOOH_{(s)}$] are highly metal-reactive and are typically widespread throughout most environmental systems, making them a dominant solid phase controlling metal behavior in the environment (Benjamin et al. 1980). Oxyhydroxides of Fe and Mn are redox-sensitive, with the more oxidized form of both elements, Fe (III) and Mn (III, IV), forming a solid phase, while the reduced forms, Fe (II) and Mn (II), are soluble. Thus, changes in redox conditions will profoundly affect the concentrations of these solids and associated metal behavior. Although it should be noted that reductive dissolution of both oxyhydroxides proceeds much more quickly with microbial catalysis than without and that, while abiotic oxidation of Fe (II) to Fe (III) and associated hydrolysis and formation of FeOOH(s) solid particles occurs spontaneously above pH 3 in the presence of oxygen, abiotic Mn (II) oxidation is extremely slow below pH 9 (e.g. months to years; Morel & Hering 1993, Stumm & Morgan 1996).

Natural organic matter (NOM), both living (e.g. micro-organisms) and dead (both labile and refractory, e.g. humic and fulvic acids), is an efficient, but often dynamically reversible, metal sequestration pool (Gustafsson et al. 2003, Smiejan et al. 2003, Jacob & Otte 2004, Boullemant et al. 2004). Sequestration by NOM can be especially important in systems of high NOM concentration such as wetlands and eutrophic lakes, typically through sorption reactions associated with their highly electrically charged surfaces. Decomposition of organic matter often leads to release of associated metals into solution and thus the ability of the NOM pool in any given system to hold onto metals will reflect the relative magnitude of decompositional processes.

In the Akaki Rivers and Aba Samuel Lake, the high organic matter content of the suspended particulate matter is expected to have significant impacts on the speciation and bioavailability of heavy metal pollutants and any subsequent remediation plan. Heavy metal pollutants such as Pb and Ni are typically associated with the organic matter. If this is the case, the bioavailability and toxicity of the metals will be largely determined by complexion with sites on organic matter.

As interactions of dissolved metals with particles are important in their geochemical cycles (Morel et al. 1991), and both dissolved and particulate trace metal data are available for the Akaki Sub-Basin, the distribution coefficients (K_d) can be calculated as:

$$K_d = C_{SPM} / C_{dissolved}$$

Where, K_d is the distribution coefficient, C_{SPM} is the concentration of metal bound to SPM and, $C_{dissolved}$ is the dissolved concentration of the metal.

The calculated K_d values for the heavy metal ions indicated that higher proportion of Pb, As, Ni, Zn, and Cu are bound to SPM in both surface water and groundwater. For most of the contaminated surface water and groundwater log K_d values of As, Ni, Cu, and Pb are between 5 and 6 l/kg while about 4 l/kg is determined for Zn. These K_d values are well correlated to the ranges of K_d values (l/kg) compiled from literature for Ni (3.5 - 5.7 l/kg), Pb (3.4 - 6.5 l/kg), As (2.0 - 6.0

l/kg), Cu (3.1 - 6.1 l/kg), and Zn (3.5 - 6.9 l/kg) by Allison et al. (2005). The K_d values calculated for Fe and Mn are so high that comparison with the other heavy metals is not considered. It might be possible to conclude that most heavy metals in the Aba Samuel Lake and the surrounding water sources are associated to the solid phase than the dissolved phase. The direct utilization of these surface water and groundwater without any treatment makes the toxic heavy metals bio - available through ingestion and skin contact. Therefore, the high concentrations of these heavy metals on the particulate matter should be considered as a potential health hazard to the communities that rely on such water sources.

The evaluation of metal pollution is an important aspect of most water quality assessment programmes. In the Akaki Sub-Basin except some heavy metals such as Fe, Mn, Zn, and the radioactive element U, most of the dissolved concentrations of heavy metals are below and close to the US EPA MCL. Elevated concentrations of Fe and Mn exceeding the EPA MCL of 0.3 mg/l, 0.05 mg/l respectively are measured in both groundwater and surface water samples. The dissolved concentrations of Zn and U are also at critical values compared to the EPA MCL of 5 mg/l and 30 μg/l respectively. Weathering of volcanic rocks and release from industrial effluents are major sources of Fe and Mn in the Akaki waters. Since Zn is found in the earth's crust in considerable amount, weathering processes can contribute significant amount of Zn in natural water. The United States Environmental Protection Agency (US EPA) considers eight trace elements as high priority: As, Cd, Cu, Cr, Pb, Hg, Ni and Zn of which Cu, Zn, Ni, Pb and As are of high importance in this study.

The absence of Fe and Mn in some priority lists results from their frequent classification as major elements. In aqueous solution, the occurrence of Fe is dependent on environmental conditions, especially oxidation and reduction. Flowing surface water, that is fully aerated, should not contain more than a few micrograms per liter of uncomplexed-dissolved Fe at equilibrium in the pH range 6.6 to 8.5. However, this value is much higher in surface water samples of Akaki.

In the Akaki River, the dissolved concentration of Fe is relatively higher than that of groundwater. In groundwater, however, much higher levels can occur. In anoxic groundwater with a pH of 6 to 8, Fe^{2+} concentrations can be as high as 50 mg/l and concentrations of 1 to 10 mg/l are common. The Fe originates by solution at sites of either reduction of ferric hydroxides or oxidation of FeS (Hem 1989) and the process is strongly influenced by microbiological activity.

The relatively higher oxidizing condition in most groundwater and neutral pH (7- 8) cause the precipitation of $Fe(OH)_3$. Hence, the total concentration of Fe in groundwater is low. On the other hand, the suboxic to anoxic conditions prevailing in most surface water of the Akaki Sub-Basin (DO less than 3 mg/l) cause the release of Fe and Mn in to surface water. In contrary to this fact, concentrations of Fe and Mn in Gerado pool at Gr7 is around 380 µg/l though the prevailing redox condition is oxic (DO=8.5 mg/l). This could be attributed to the high evapotranspiration rates that cause the concentration of these metals derived from the rock-water interaction to be enriched. The Gerado pool water is situated at 2172 m elevated hill that is recharged by rainwater. There is no anthropogenic influence in this pool, which is located relatively far from the main polluting rivers of Akaki.

In comparison to Fe, Mn is found in high concentrations in surface water of the Akaki Sub-Basin, the highest being in Aba Samuel Lake (Dk4). Based on comparison to typical background levels of Mn in surface water or groundwater it seems likely that some groundwater and surface water samples where Mn is detected contain only natural background levels. Although ambient Mn levels are about 200 µg/l in a number of cases, high levels (in excess of 1,000 µg/l) have been detected in most surface waters indicating that Mn containing wastes might play significant role in the contamination of surface water. The transport and partitioning of Mn in water is controlled by the solubility of the specific chemical form present, which in turn is determined by pH, Eh (oxidation-reduction potential), and the characteristics of the available anions. In the Akaki Sub-Basin Mn (II) predominates in most waters (pH 7) but may become oxidized at higher pH values greater than 8. In relatively oxidized water, the solubility of Mn(II) may be controlled by Mn oxide equilibria (Ponnamperuma et al. 1969), with Mn being converted to the Mn(II) or Mn(IV) oxidation states (Rai et al. 1986). In extremely reduced water, the fate of Mn tends to be controlled by formation of a poorly soluble S^{2-}. In surface and groundwater of the Akaki Sub-Basin, the concentrations of Mn in dissolved as well as SPM phase are higher than those of Fe.

Zinc

One of the heavy metals occurring at the highest dissolved concentrations in the Akaki water is Zn. In surface water samples the dissolved fractions of heavy metals are in the following order from highest to lowest: Zn > Ni > Cu > As > Pb while in groundwater the order is somewhat different, Zn > Cu > Pb > Ni > As. The mean concentrations of Zn recorded in wet seasons are relatively higher at samples sites E10, Ju11, Tl12, Iv13, Ls16, and Ak17 than those in the dry season (Appendix 2). Three metals have been found at critical values compared to water quality criteria set by the US EPA: dissolved Zn, Pb, and Ni. In groundwater, concentrations of these metals reach 3200 µg/l, and 7.44 µg/l for Zn and Pb respectively. On the other hand, concentrations of Zn bound to suspended particulate matter in Lake Aba Samuel (Dk4) (4000 µg/g) are ten times higher than Cu and Pb concentrations. In the Lake outlet (Dk5) 250 µg/g of Zn is measured. The concentration of Zn bound to SPM in Big Akaki River at Ku1 is 630 µg/g as compared to the lower 70 µg/g in Small Akaki River at Ju11. According to the LAWA classification, Lake Aba Samuel falls in LAWA class IV indicating that the lake is very strongly contaminated with Zn. In groundwater samples, Zn bound to SPM/colloidal is much higher than that of the surface water. The highest concentrations of Zn bound to SPM/colloids measured are 13000 µg/g, 14000 µg/g, 15000 µg/g, and 38000 µg/g at sample sites Ak17, De6, Dk3, and E10 respectively. Similar to Cu, higher content of Zn bound to SPM in groundwater might be due to the higher association of Zn with fine fractions of SPM and colloids. Based on the LAWA classification for Zn most of the groundwater falls in the very strongly contaminated category (class IV). Soil-water interactions are the main sources of higher concentrations of Zn bound to SPM in the Akaki groundwater. In natural waters, Zn species are limited to Zn hydroxides, Zn silicates, Zn carbonates, and the free ion Zn^{2+} (Novotny 2002). The noncomplexed Zn^{2+} ion dominates at pH 8 and uncharged $ZnCO_3$ at higher pH. Hydrolysis becomes significant at a pH>7.5, although the hydroxy complexes $ZnOH^+$ and $Zn(OH)_2$ would not dominate in the presence of carbonate at groundwater concentration levels (Salbu 1995). Except Zn carbonates, the other forms of Zn are expected to be present in the Akaki waters at the prevailing redox conditions. The solubility of Zn begins to decrease around a pH of 6 - 7 (Bethke 1996, Holmstrom et al. 1999).

Heavy metal analysis on some rock samples taken from upstream part of Addis Ababa showed elevated concentrations of Zn. The concentrations of Zn in rock samples taken from northern part of Addis Ababa (6Kilo a, 6Kilo b, and Entoto) reached 80 ppm, 361 ppm, and 125 ppm respectively. These areas of Addis Ababa are drained by tributaries of the Big Akaki River. On the other hand, rock samples taken from parts of Addis Ababa that are drained by tributaries of the Small Akaki River at sample sites Gojam Berenda and Asmara Road showed Zn concentrations up to 530 ppb and 217 ppm respectively (Alemayehu 2001). Weathering of these rocks can mobilize Zn in surface water and groundwater. Very high abundances of soluble Zn are present under well-oxidized conditions and at pH 5 to 6.5, whereas low abundances of soluble Zn are present at pH 8 under all redox conditions and at pH 5 to 6.5 under moderately and strongly reducing conditions (Gambrell et al. 1991). Zinc concentrations in surface and groundwater are significant under the prevailing pH condition of the Akaki waters (pH 7 - 8).

Lead

Dissolved concentrations of Pb remained within a relatively small range in almost all surface water samples. However, in groundwater samples Pb has the most variable concentrations ranging from 0.08 µg/l at sample site Ls16 to 7.44 µg/l at

De2. The values recorded in E10 and T12 groundwater samples for dissolved Pb 2.5 µg/l and 4.7 µg/l respectively are also high relative to other samples from this area. For the most part, the dissolved concentrations of Pb in groundwater samples are higher than that of the surface water samples. In surface water samples, Pb is well below the US EPA MCL guidelines (15 µg/l). Concentrations of Pb bound to suspended particulate matter in Lake Aba Samuel (Dk4) reach 267 µg/g while in the Lake outlet (Dk5) 20 µg/g is measured. The mean concentrations of Pb recorded in wet seasons have been relatively higher at samples sites E10, Tl12, Iv13, Ls16, and Ak17 than those in the dry season (Appendix 2). According to the LAWA classification Lake Aba Samuel falls in LAWA class III indicating that the lake is strongly contaminated with Pb. Lead bound to SPM/colloids in groundwater samples is much higher than that of the surface water. Based on the LAWA classification Pb contamination of groundwater falls in the moderately to strongly contaminated (class II - III) in 46 % of the samples. The highest concentration of Pb (1300 µg/g) bound to SPM are detected at sample site De2 exceeded class IV of LAWA classification indicating a very strong contamination. Rock-water interactions are the main sources of Pb in the Akaki groundwater. The concentrations of Pb in rock samples taken from upstream part of the Akaki catchment at sample site 6Kilo a, 6Kilo b, and Entoto reach 24 ppm, 23 ppm, and 1 ppm respectively (Alemayehu 2001). Tributaries of the Big Akaki River drain these parts of Addis Ababa. On the other hand, rock samples taken from parts of Addis Ababa that are drained by tributaries of the Small Akaki River showed Pb concentrations up to 15 ppm and 18 ppm at sites Gojam Berenda and Asmara Road respectively. Other than geogenic sources, some battery manufacturers in Addis Ababa contribute to the anthropogenic sources of Pb in the Akaki water. Lead is tightly bound under strongly reducing conditions by S^{2-} mineral precipitation and complexation with insoluble organic matter, and is very effectively immobilized by precipitated Fe oxide minerals under well-oxidized conditions (Gambrell et al. 1991). Lead adsorption peaks at approximately pH 5, and then precipitates as Pb carbonate as pH increases.

Nickel

Nickel is an element of environmental concern. It belongs to the suite of toxic heavy metals (Ni, Cu, Zn, Cd, Pb, and Hg) that are subject to concentration limits in drinking water (Office of Environmental Health Hazard Assessment, 2001). The mean concentrations of dissolved Ni recorded in wet seasons are relatively higher at samples sites Ju11, Iv13, and Ak14 than those in the dry season (Appendix 2). Recharge of groundwater by surface runoff and the interaction between groundwater and the Akaki River could be the possible reasons for the relatively elevated heavy metal concentrations at these sample sites. In contrary to Pb, the dissolved concentrations of Ni in the Akaki surface water are ten times higher than that of groundwater samples. In comparison to the low groundwater concentration of 1.7 µg/l at Iv13 the concentration of Ni at Gerado pool (Gr7)reaches up to 45 µg/l. Background rock Ni concentrations in the upstream part of Addis Ababa reach up to 27 ppb, 17 ppb, and 8 ppb Ni in the localities mentioned earlier (Alemayehu 2001). Compared to content of Ni in natural uncontaminated soils (40 µg/g) according to Lindsay (1979) the Akaki soils are enriched with Ni (46 - 60 µg/g). Nickel is a relatively mobile heavy metal. In natural waters, Ni is transported in both particulate and dissolved forms. The pH, oxidation - reduction potential, ionic strength, type, and concentration of organic and inorganic ligands (in particular, humic and fulvic acids), and the presence of solid surfaces for adsorption (in particular, hydrous Fe and Mn oxides) can all affect the transport, fate, and biological availability of Ni in fresh water (Callahan et al. 1979). In the Akaki waters most of the Ni is associated with the suspended particulate matter, the high calculated K_d values (1.6 x 10^6 l/kg) e.g. in Aba Samuel Lake (Dk4) confirms the higher association of Ni to SPM. Under aerobic conditions and pH < 9, the compounds that Ni forms with hydroxide, carbonate, sulphate, and naturally occurring organic ligands are sufficiently soluble to maintain aqueous Ni^{2+} concentrations above 60 µg/L (Callahan et al. 1979).

The concentrations of Ni in rock samples taken from upstream part of the Akaki catchment at sample site 6Kilo a, 6Kilo b, and Entoto reach 27 ppm, 17 ppm, and 8 ppm respectively (Alemayehu 2001). These parts of Addis Ababa are drained by tributaries of the Big Akaki River. On the other hand, rock samples taken from parts of Addis Ababa that are drained by tributaries of the Small Akaki River showed Ni concentrations up to 54 ppm and 10 ppm at sites Gojam Berenda and Asmara Road respectively. In uncontaminated Vertisols and Fluvisols (0 - 20 cm) of Akaki the mean concentration of Ni reach 50 ppm and 6 ppm respectively showing that the values markedly surpassed the reported average soil contents for Ni (Angelone and Bini 1992, McGrath 1995, Adriano 2001, Pendias and Pendias 2001) of the world. Erosion and weathering activities may induce Ni in to surface and groundwater systems.

Arsenic

More than one-half of the surface water and groundwater samples collected had dissolved As concentrations well below 2 µg/l, the lowest being measured in the groundwater samples. The principal tributaries generally contained As concentrations very similar to that of the main Akaki River. The highest concentrations of As in surface water are measured in Aba Samuel Lake at Dk4 (2.8 µg/l) and Small Akaki River at El18 (2.13 µg/l). In general, the dissolved concentrations of As in groundwater samples are below that of the surface water samples. However, these concentrations decreased downstream from the confluences through simple dilution, precipitation, or adsorption onto sediment particles. In groundwater samples the mean concentrations of dissolved As recorded in wet seasons are relatively lower than that of the dry season (Appendix 2). Recharge of groundwater by surface runoff and the interaction between groundwater and the Akaki River could be the possible reasons for the relatively elevated heavy metal concentrations at these sample sites. The relatively low concentrations of dissolved metals in the surface water especially in the Aba Samuel Lake and drainages downstream of the Akaki River are due to the high pH, which acts to reduce metals solubility. Iron and Al precipitate directly, forming hydrous oxides that are nearly insoluble at neutral pH. Coprecipitation of other metals is attenuated by adsorption onto the surface of the precipitating minerals.

Arsenic, for example, adsorbs over a wide range of pH conditions down to 4 or less.

The concentrations of As in rock samples taken from upstream part of the Akaki catchment at sample site 6Kilo a, 6Kilo b, and Entoto reach 2 ppm, 4 ppm, and 1 ppm respectively (Alemayehu 2001). These are parts of Addis Ababa that are drained by tributaries of the Big Akaki River. On the other hand, rock samples taken from parts of Addis Ababa that are drained by tributaries of the Small Akaki River showed As concentrations up to 2 ppm and 7ppm at sites Gojam Berenda and Asmara Road respectively. In uncontaminated Vertisols and Fluvisols (0 - 20 cm) of Akaki the mean concentration of As reach 3 ppm and 1 ppm respectively indicating that the As contents are lower than the reported average soil contents for As (Adriano 2001) of the world. In addition to the geogenic inputs, agricultural fields at the suburbs of Addis Ababa contributed to the higher concentrations of As bound to SPM in surface and groundwater.

Copper

Dissolved Cu in the Akaki River originates mostly from industrial and municipal wastewaters released. The dissolved concentrations of Cu in groundwater are generally lower as compared to the surface water samples. For those groundwater wells located close to the Akaki River such as Iv13 and E10, concentrations raised to 5 µg/l and 3.5 µg/l respectively. The mean concentrations of dissolved Cu recorded in wet seasons are relatively higher at samples sites E10, Ju11, and Iv13 than those in the dry season (Appendix 2). Recharge of groundwater by surface runoff and the interaction between groundwater and the Akaki River could be the possible reasons for the relatively elevated heavy metal concentrations at these sample sites. These relatively elevated values could be caused by pollution effect of the Akaki River. In the absence of dramatic change in soil properties and pollution, groundwater concentrations can be expected to be similar to adjacent bodies of surface water. Copper concentrations measured in the Nile River, as well as in the groundwater and drinking water (post-filtration river water) near the Aswan Dam were similar in all three media, between 5 - 6 µg/l (Ismail 1996). These are in agreement with concentrations measured in the Akaki groundwater samples (3.5 - 5 µg/l). Rock - water interactions are the main sources of Cu in the Akaki groundwater. The solubility of Cu begins to decrease around a pH of 6 - 7 (Bethke 1996, Holmstrom et al. 1999). In the Akaki surface water concentrations of dissolved Cu generally decrease in the downstream direction, especially near the confluence of the Akaki River and Aba Samuel Lake. Concentrations of Cu bound to suspended particulate matter in Lake Aba Samuel (Dk4) reach 434 µg/g that is twice higher than that of Pb. In the Lake outlet (Dk5) 20 µg/g of Cu is measured that is similar to Pb. The concentration of Cu bound to SPM in Big Akaki River at Ku1 is 40 µg/g as compared to the lower 5 µg/g in Small Akaki River at Ju11 and El18. According to the LAWA classification, Lake Aba Samuel falls in LAWA class III - IV indicating that the lake is strongly to very strongly contaminated with Cu. In groundwater samples Cu bound to SPM/colloids is much higher than that of the surface water. This might be due to the strong association of Cu with fine fraction of SPM / colloids in groundwater in higher proportion. Based on the LAWA classification for Cu most of the groundwater falls in very strongly contaminated (class IV). The highest concentration (5846 µg/g) of Cu bound to SPM detected at sample site E10 exceeded class IV of LAWA indicating a very strong contamination.

On the other hand, concentrations of Cu in rock samples taken from upstream part of the Akaki catchment at sample site 6Kilo a, 6Kilo b, and Entoto reach 25 ppm, 25 ppm, and 19 ppm respectively (Alemayehu, 2001). These areas are parts of Addis Ababa that are drained by tributaries of the Big Akaki River. Rock samples taken from parts of Addis Ababa that are drained by tributaries of the Small Akaki River showed Cu concentrations up to 10 ppm and 21 ppm at sites Gojam Berenda and Asmara Road respectively. In uncontaminated Vertisols and Fluvisols (0 - 20 cm) of Akaki the mean concentration of Cu reach 25 ppm and 6 ppm respectively indicating that the As contents were lower than the reported average soil contents for Cu (Baker and Senft 1995, Adriano 2001) of the world. Copper leaching from soil affects groundwater concentrations, due to Cu concentrations in runoff water that eventually percolates through the soil. The median concentration of Cu in natural water is 4 - 10 ppb. It is predominantly in the Cu(II) state. Most of it is complexed or strongly bound to organic matter; little is present in the free (hydrated) or readily exchangeable form. The combined processes of complexation, adsorption, and precipitation control the level of free Cu(II). The chemical conditions in most natural water are such that, even at relatively high Cu concentrations, these processes will reduce the free Cu(II) concentration to extremely low values. Sediment is an important sink and reservoir for Cu. In relatively clean sediment, the Cu concentration is <50 ppm; polluted sediment may contain several thousand ppm of Cu. The form of Cu in sediment will also be site-specific. Organics (humic substances) and Fe oxides are the most important contributor to binding of Cu by aerobic sediments. In anaerobic sediment, Cu(II) will be reduced to Cu(I) and insoluble cuprous salts will be formed.

6.3 Efficiency of Filter Media for Heavy Metal Removal

The results of XRD, XRF, and sequential extractions of the filter media reveals that pyroclastic fall deposits, scoria, Mn-rich fluvial deposits and laterite are potential filter materials in removing heavy metals from polluted surface and ground waters (Table 4.3). The high Fe-oxide (hematite) content (10 - 20%) and the permeability (1.9×10^{-3} m/s) of DQ scoria make it a potential filter material. EBF laterite with a slightly lower permeability (9.4×10^{-5} m/s) and high content of montmorillonite (12%) and kaolin (5%) is the other potential filter media that can be applied in water purification. KVA volcanic ashes with 5% of clay fraction and permeability of 2.5×10^{-5} m/s can also be considered as potential filter media.

Soils formed on volcanic ash deposits contain structurally disordered alluminosilicates and allophane (Sposito 1989). The structure of allophane is thought to consist of a 1:1 aluminosilicate layer framework riddled with defects (vacant ion sites) and containing Al in both the tetrahedral and the octahedral sheets (Sposito 1989). In this case surface charges are developed either through the desorption and adsorption of protons in 1:1 clay minerals and "amorphous materials", including allophane, imogolite, and oxyhydroxides of Fe and Al (Sposito et al. 1984). This favors sorption of cations from solutions. The composition of allophane ranges from that of halloysite (Al/Si = about 1) to an Al - rich composition with an Al/Si ratio of about 2 (Buol et al 1997). KVA volcanic ash has been formed on a thick deposit of volcanic material that might partly be transformed to volcanic soils. EBF laterite is also derived from basic volcanic materials that are rich in ferromagnesian minerals. These common structures of allophane may also partly present in KVA volcanic ashes. Bleeker et al. (1990) indicated that clay mineralogy of oxisols is strongly related to the composition of parent material. In one profile developed on ultrabasic rocks the clay mineralogy is dominated by goethite and the other, developed on basaltic rocks, have kaolinite as the major clay mineral. In contrast, the clay minerals of the soils developed on volcanic ash deposits are dominated by hydroxy Al-interlayered vermiculite, which, in one soil, is most likely formed as an alteration product of biotite.

Birnessite is one of the most important sorbing minerals for heavy metals dominantly found in MnR sediment. It represents 34% of the minerals composing MnR sediment. The basic building block of MnOx minerals is the MnO_6 octahedron. These can share edges to form sheets, and corners to form tunnel-like structures. Depending on the number of octahedra on each side of the tunnel, α MnOx structure can be 1x1, 2x3, etc., where each unit refers to one octahedron. The MnOx minerals are very fine-grained and poorly crystalline which result in a very high surface area (Post 1999). They possess very high adsorption capacity and selectively remove large amounts of heavy metals from these surroundings. Manganese oxide minerals are highly charged, with the point of zero charge (pH_{ZPC}) of 2 - 4.5. Mn (hydr)oxides are very reactive components of natural environments and affect the fate, transport, and bioavailability of a variety of heavy metals and organic compounds (Davis et al. 1990). Several authors have noted that birnessite readily participates in cation exchange reactions (Ching and Suib, 1997; Post, 1992; Brock et al., 1998). Most of the adsorption of metal ions occurs at the surface of the MnOx minerals due to unsatisfied oxygen or hydroxyl bonds and only a minor amount enters the layers in between sheets (Taylor 1986). A systematic study, based on natural and synthetic species of Mn- and Fe-oxides in two background electrolytes showed that Mn - oxides are generally more efficient sorbents of Pb than Fe-oxides (O'Reilly 2000).

Hematite is the other important mineral constituting DQ scoria. Hematite has red color an indication of oxidation of Fe(II) to Fe(III) and is often found in nature, particularly in tropical or subtropical soils. Like goethite, hematite is extremely stable and, hence, often the end member of transformations of other Fe oxides. Hematite can form in several ways but one common way implies an internal rearrangement and dehydration of ferrihydrite (Schwertmann et al. 1991). Foreign cations like Al, Cu, Cr, and Mn can substitute for Fe in hematite (Lim-Nunez et al. 1987, Singh et al. 1992, Cornell and Schwertmann 1996). Compared to goethite hematite has low sorption capacity. Selectivity sequence of heavy - metal cation adsorption has been found for hematite in the order Pb > Cu > Zn > Cd > Ni > Mn (Schwertmann et al, 1989).

Montmorillonite, from a structural standpoint, is a 2:1 phyllosilicate and therefore possesses two tetrahedral sheets apart from the octahedral sheet (Fig. 6.4; Güven 1988). The negative layer charge resulting from isomorphic substitutions is balanced by the sorption of exchangeable cations in interlayer sites (Sposito 1984). The uptake kinetics of cation exchange is fast, and exchangeable interlayer cations can be replaced by solute ions by varying the concentration of the aqueous ions (McBride et al. 1975, Sposito 1984, Tang et al. 1993, Chisholm - Brause et al. 1994, Verburg et al. 1995, Papelis et al. 1996, Muller et al. 1997, Schlegel et al. 1999b). Montmorillonite is an important mineral in EBF laterite, which contributes to the removal of heavy metals from the synthetic polluted water. The sorption of Ni,

Cu, Zn, As and Pb by montmorillonite has been discussed in (Baeyens, et al. 1995). Na-montmorillonite has been used as cost effective sorbents in the removal of heavy metals from industrial wastes containing elevated amounts of heavy metals. In this study the simulated polluted water has considerable amount of organic matter such as peptone (160 mg/l), meat extract (110 mg/l), and urea (30 mg/l). In addition to the spiked montmorillonite, the movement of the synthetic polluted water creates a medium where heavy metals will be removed on the clay organic matter complex biofilm.

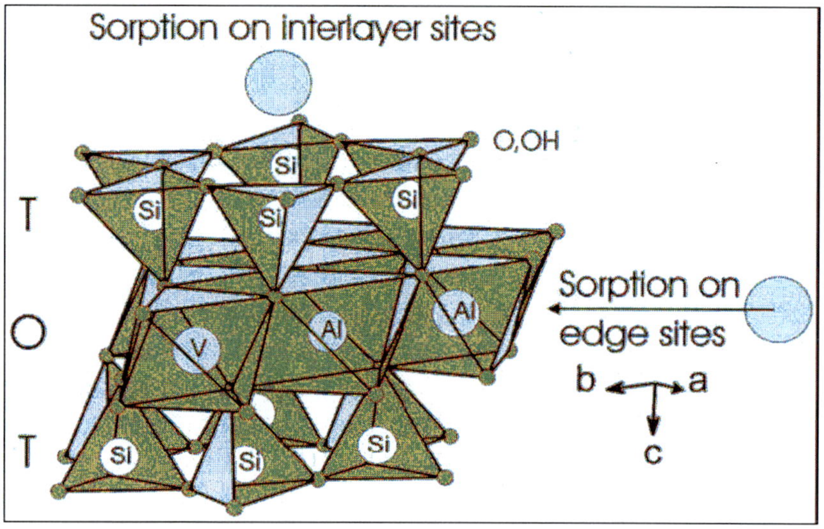

Fig. 6.4. Uptake modes on dioctahedral aluminous clays (V = vacancy). Structure of montmorillonite after Tsipursky and Drits (1984).

As kaolin represents a considerable fraction of EBF laterite and KVA volcanic ash it contributes to the retention of heavy metals from the simulated polluted water. Kaolin is a representative layered alumino-silicate mineral with the structure of the tetrahedral (Si centre) and octahedral (Al centre) sheet in a 1:1 ratio. The permanent structural charge of kaolinite is minor, hence cation adsorption takes place mainly at the proton bearing surface functional group such as silanols and aluminols exposed at the edge of the sheets.

Estimated apparent removal rates for each metal from the synthetic polluted water are different for the particular metal tested. Metal sorption on kaolinite and illite does not differ much (Lackovic et al. 2004) and it is known to be a rapid reaction since their exchange capacities are mainly due to external surface and edge sites readily accessible to cation exchange (Jardine et al. 1984). In fact, cation exchange on clays without narrow interlayer space such as kaolinite appears to be instantaneous in comparison to exchange on smectite which can be related to its freely expanding interlayer space (Jardine et al. 1984) and limited by the rate of cation diffusion through this region (McBride 1994).

As the dominant mineral in volcanic materials, quartz has physical purifying capacity in removing suspended particulate matter and some organic matter. Most slow sand filters work through the formation of a gelatinous layer (or biofilm) called Schmutzdecke in the top few millimeters of the fine sand layer. The use of filter media constituted of quartz sand and volcanic materials are often an appropriate technology for poor and isolated areas. Water treatment using these filter media require little or no mechanical power, chemicals or replaceable parts, and they require minimal operator training and only periodic maintenance.

Batch experiments using the spiked Na-rich Wyoming type montmorillonite as adsorbent in a synthetic polluted water background solution reveal significant removal capacities for Cu, Zn and Pb. High sorption affinity towards montmorillonite is observed in this experiment for Pb (94%) while the affinity decreases for Cu (51%), Ni (36%), Zn (26%) and As (8%). The high SPM concentration (1 g/l) and dissolved organic matter content (0.081 mg/l) derived from the synthetic polluted water played significant roles in sorption of metals within the columns. Heavy metal cations sorb to soil organic matter (SOM) and other forms of humified natural organic matter (NOM). Soil organic matter is an important component of the soil since it has a high surface area, and has functional groups that are Lewis bases (e.g., carboxyl and phenol groups) that metals can form chemical bonds with (Sparks 1995). There are factors that affect NOM

complexing of metals such as pH, ionic strength, NOM source characteristics and competing ligands. Humic substances can be important metal complexing agents and may also significantly coat the solid mineral controlling the electrochemistry of the solid phase (Sparks 1995). A detailed analysis of the adsorption behavior of several inorganic and organic complexing ligands and the effects of these ligands on trace metal uptake by amorphous Fe oxide has been reported in Leckie (1978).
It is anticipated that the suspended solid surfaces in these column experiments have been coated by adsorbing organic matter that were derived from the SPW, which latter contribute in removing certain heavy metals such as Pb and Cu significantly. There are evidences that have shown the influence of organic matter in removing some heavy metals from SPW. The low dissolved oxygen content (~ 2.3 mg/l) and foul odor of the spent media are indications of reducing condition in the columns. Observation of the color and smell of the column sludge shows that formation of H_2S is also likely, indicating the role of organic matter and reducing condition formed within the columns. These hydrochemical characteristics are typical of the Akaki waters, which in most parts of the year show reducing conditions, low dissolved oxygen and BOD amount.

6.3.1 Comparison of Removal Efficiencies of Columns

Considering the simulated polluted water multiple-metal-removal experiments, several important trends can be identified. For QTZ sand, it appears that retention of Ni and Zn is similar and relatively weak while that of Pb and Cu is as strong as the other filter media. Retention increased for EBF laterite (Table 6.4), with total amount removed descending in the following order: Pb ≥ Zn > Cu > Ni >> As. The greater retention of metals to the EBF laterite compared to KVA volcanic ash and QTZ sand suggests that the EBF laterite is a better filter media than the other media collected from Ethiopia. Similarly removal of heavy metals also increased in columns filled with DQ scoria, with percent sorbed descending in the following order: Pb ≥ Zn > Cu > Ni >> As (Table 6.4). In MnR sediment the removal of heavy metals and As is the highest compared to all other filter media. Particularly the removal of As is strong compared to DQ scoria and EBF laterite. Therefore,

MnR sediment is the best performing media in removing all heavy metals and As from the synthetic polluted water. In addition to the spiked montmorillonite MnR sediment contains substantial amount of Mn oxide mineral (birnessite) that played greater role in the removal of heavy metals and As.

The two adsorption sequences mentioned here for EBF laterite and DQ scoria did not exactly follow the order of the electronegativity of the metal cations, which is, according to Evans (1966), Cu (1.9), Pb (1.8), Ni (1.8), and Zn (1.6). However, the sequences are in agreement with adsorption sequences indicated for tropical soils (Veeresh et al. 2003).

Table 6.4. Apparent removal efficiencies of the filter media (%).

Samples	Ni	Cu	Zn	Pb	As
EBF laterite	92.4	97.4	98.6	94.2	56.0
DQ scoria	74.0	94.7	97.6	96.9	60.9
KVA volcanic ash	81.5	94.8	97.7	97.2	-17.8[1]
PYC pyroclastics	79.6	91.2	94.6	98.2	-13.9[1]
QTZ sand	72.8	94.4	98.3	96.1	-24.3[1]
MnR sediment	99.0	99.1	99.3	98.5	90.5

[1]Negative removal values represent release of the element from the column pack.

The cumulative effect of these sorbing minerals in MnR sediment resulted in the efficient removal of all heavy metals and As. Unlike QTZ sand and PYC pyroclastics, EBF laterite contains considerable amounts of clay minerals (kaolinite and montmorillonite) in addition to the spiked montmorillonite. This in turn contributes to polar (charged) surfaces that can effectively sorb metals or charged metal - hydroxides, and this feature of the filter media likely explains its improved performance relative to QTZ sand and PYC pyroclastics. The best metal removal from solution is observed for column filled with MnR sediment with percent sorbed descending in the following order: Zn > Pb > Cu > Ni > As (Table 6.4). The strong performance of MnR sediment may be attributable to its relatively large internal porosity and correspondingly charged surface area. Substantial amount of hematite in DQ scoria also contributed for the removal of heavy metals and As from solutions from column filled with DQ scoria. Regarding the filter media collected from Ethiopia a general removal capacity trend could be established for Zn, Cu, Pb, Ni, and As: DQ scoria ≥ EBF laterite > KVA volcanic ash > QTZ sand. However, the MnR sediment is the best performing media compared to all other media while the PYC pyroclastics is less performing particularly in removing Ni and As.

Though, there is no remarkable difference in the removal of Pb, Cu, and Zn between the filter media the results obtained in the batch experiment are in agreement with those reported in Elliot et al. (1986), Basta et al. (1992), Echeverria et al. (1998), which showed that Pb and Cu metal adsorption is much greater than Cd, Ni and Zn adsorption. The adsorption of Cu, Zn, Ni, As and Pb on Na-montmorillonite in simulated polluted water background solution followed the following sequence: Pb > Cu > Ni > Zn > As. For comparison purposes the predicted affinity sequences of metals based on their properties-ionic (Huheey 1972), hydrolysis constant (Baes et al. 1976) and softness (Misono et al. 1967) are shown in table 6.5.

In the absence of competition, the observed greater adsorption affinity of Pb than that of Cu, Ni, Zn and As would be related to some relevant metal properties. For electrostatic adsorption of metals with equal charge (Z) on ion-exchange materials, metal affinity should be inversely related to unhydrated radii (r). If the metal adsorption on the clays is entirely electrostatic, ions of higher ionic potentials (Z^2/r) should be adsorbed more strongly.

Table 6.5. Affinity sequences of divalent metals on the basis of several metal properties. (from Basta and Tabatabi 1992).

Metal property	Metal affinity sequence
Ionic potential	Ni > Cu > Zn > Cd > Pb
	(5.71) (5.48) (5.33) (4.21) (3.39)
Hydrolysis constant	Pb > Cu > Zn > Ni > Cd
	(7.8) (8.0) (9.0) (9.9) (10.1)
Softness	Pb > Cd > Cu > Ni > Zn
	(3.58) (3.04) (2.89) (2.82) (2.34)

The softness concept is derived from the hard-soft acid base (HSAB) theory of Pearson (1963), in which hard acids tend to associate with hard bases and soft acids with soft bases. Misono et al. (1967) derived a softness parameter that numerically expressed the tendency of a metal to form covalent bonds (Table 6.5). Affinity sequences reported in table 6.5 show that the HSAB does predict the metal affinity sequence observed (except for the order of Cu and Ni) for Na-montmorillonite in simulated polluted water background solution. Basta and Tabatabai (1992) also reported a good agreement between the observed metal affinity sequence (for Pb, Cu, Ni, Zn, and Cd on soils) with that predicted by HSAB theory except for the order of Cd and Zn.

The degree to which metals are hydrolyzed, which is represented by the hydrolysis constant, is likely to be a major factor determining the amount of metal retained at any given pH. Reactions between metal ions and water to form hydrolysis products (i.e., $MeOH^{+}$) are common to most metals and can be expressed in general form (Baes and Mesmer 1976):

$$xM^{z+} + yH_2O = M_x(OH)^{(xy-z)+} + yH^+$$

The affinity sequence based on the first hydrolysis constants (formation of $MeOH^{+}$ complexes) is reported in table 6.5. Metal affinity sequences consistent with the adsorption of $MeOH^+$ species have been used to describe metal adsorption by soils and Fe and Al oxides (Kinniburgh and Jackson 1981, McBride 1989). The high affinity of Pb can be predicted based on metal hydrolysis regardless of the adsorbent type.

6.4 Heavy Metal and SPM Removal from Polluted Water by Filter Media

The column tests are conducted using a complex waste solution representing the natural hydrochemical conditions existing in the Akaki Sub-Basin. The retention of suspended particulate matters (SPM) within the test columns demonstrated that natural waters of high SPM concentrations could be purified by the filter media tested. The adsorption of trace metals onto SPM is influenced by the geochemistry of the SPM including its clay mineral, Fe-oxide and Mn-oxide content (Bibby et al 2005). The main constituent of SPM in this investigation is Wyoming type Na montmorillonite. Some general physical and chemical characteristics of Na montmorillonite are discussed below. In this study we gave emphasis to the distribution and behavior of the suspended particulate matters (SPM) because most of the metals are bound to SPM and they are the major controlling factors in the mobilization of heavy metals in the Akaki catchment. Three of the possible mechanisms by which trace metals can be concentrated by sediments and suspended particulate matter include physico-chemical adsorption from the water column, biological uptake particularly by bacteria and algae, and the sedimentation and physical entrapment of enriched particulate matter (Hart 1982). Strong associations of Mn, Fe, Ni, Cu, Zn, As, and Pb have been observed in the SPM samples taken from the Akaki surface water and groundwater.

6.4.1 Retention of SPM

The removals of SPM and thereby heavy metals bound to SPM are effective for all filter media tested. The column experiments also indicated that heavy metals of concern remain strongly bound to the particulates during the experimental period at the existing pH condition likely to occur in the Akaki Sub-Basin. On the other hand the filter media captured heavy metals bound to SPM from the influents over the experimental period. As expected, the behavior of the total suspended solids and particulate - bound metals paralleled each other.

The filter media are capable of providing efficient physical removal of contaminants associated with particulate matter from the synthetic polluted water. The synthetic polluted water usually moves very slowly through the filter media due to the characteristic low flow rate (0.12 ml/min) condition and the resistance provided by spiked montmorillonite that are already trapped by the filter media. Sedimentation of suspended solids is promoted by the low flow velocity and by the fact that the flow is often laminar (not turbulent) in the column experiment so as to maintain the laboratory conditions constant.

6.4.2. Removal Mechanisms of Heavy Metals from Polluted Water by Filter Media

In order to assess the reactivity of the species or binding forms of heavy metals in solid materials, various extraction media, electrolytes ($CaCl_2$ or $MgCl_2$), pH buffers of weak acids (acetic or oxalic acid), chelating agents (EDTA or DTPA), reducing agents (NH_2OH), strong acids (HCl, HNO_3, $HClO_4$, HF) or basic reagents (NaOH, Na_2CO_3) have widely been used (Lund 1990). So far, many single or sequential extraction procedures, mainly based on the Tessier procedure or its different versions, have been applied to soils and sediments to fractionate metals by using different extractants. The applicability of an improved version of the BCR Three-Stage sequential extraction procedure on a sewage sludge amended soil (CRM

483) was proven to be an effective extraction technique (Rauret et al. 1999, 2000). In this study a BCR Three-Stage sequential extraction is also conducted to obtain useful information regarding the possible retention mechanisms involved in removing heavy metals of interest from the polluted water. Other important aspects such as the fate and mobility of heavy metals in the used filter media have also been evaluated. The potential leaching of heavy metals from the used media under the environmental conditions prevailing in the Akaki Sub-Basin will also be discussed.

From the BCR extraction procedure it is possible to observe that the metal removal mechanisms mainly involved weak sorption including surface precipitation on the media and/or montmorillonite. These are represented by exchangeable fractions that include weakly adsorbed metals retained on the media surface by relatively weak electrostatic interaction, metals that can be released by ion-exchange, co precipitated with carbonates present in many types of sediment. Changes in the ionic composition, influencing adsorption-desorption reactions, or lowering of pH could cause remobilization of metals from this fraction. Exchangeable metal ions are also a measure of those heavy metals that are released most readily into the environment.

In this study the retention of metals to the filter media can be described using two surface complexation models. It is considered that the metal retention is due to the association of these metals with diffuse ion or in outer sphere complex where they are surrounded by waters of hydration and not directly bonded to the media surface. This is evident from the BCR sequential extraction results where more than 70% of the adsorbed heavy metals are released after extracting them with acid.

In column filled with MnR sediment the acid exchangeable fraction of Ni, Zn and Cu reach up to 81% indicating that these metals are retained non-specifically at the filter media surface. About 70% of Ni, Cu and Zn are also fixed in the acid exchangeable fraction in columns filled with KVA volcanic ashes and QTZ sand. The same amount of Cu, Zn, and Pb are retained in the exchangeable fraction of PYC pyroclastics while Ni fractionation is attributed both to the exchangeable and water-soluble fraction in which the latter implied that surface accumulation can be the possible removal mechanism. In columns filled with DQ scoria and EBF laterite removals of Ni, Cu, Zn, and As are also mainly associated with the exchangeable fraction (about 80%). The acid exchangeable fraction includes weakly adsorbed metals retained on the filter media surface by relatively weak electrostatic interaction and metals that can be released by ion-exchange processes. Changes in the ionic composition, influencing adsorption - desorption reactions, or lowering of pH could cause remobilization of metals from this fraction (Ure et al. 2001, Marin et al. 1997, Ahnstrom et al. 1999, Coetzee et al. 1993 Sutherland et al. 2000, Clevenger 1990).

The exchangeable species of Ni, Cu and Zn are higher compared to the Fe- and Mn-oxides bound species in all spent media (Table 5.8 and Fig. 5.25). Compared to the organically bound (oxidizable) fraction large proportions of Zn, Cu, and Ni are present in the Fe/Mn oxide bound (reducible) fractions in the spent media. The amounts of exchangeable fraction of metals are increasing in the following order Zn > Cu > Ni > As > Pb in all spent media except for samples taken from column filled with DQ scoria indicating that these metals have been weakly retained and released by ion exchange processes. It is likely that weak sorption reactions determined Ni, Cu and Zn retention in the filter materials. In the spent media taken from MnR sediment the even fractionation of As between the exchangeable and reducible phases show that removal of As involved non-specific adsorption and precipitation. In general, As, Cu, Zn, and Ni are mostly found in the acid extractable and reducible fractions (>90%). This suggests that they have been either sorbed on various solid phases or associated with Fe/Mn oxides (Tessier et al. 1979, Hickey and Kittrick 1984, Kim and Fergusson 1991).

A small amount of Pb, compared to the amount released in the reducible and oxidizable fraction, is released from the exchangeable phase in column packed with MnR sediment indicating that removal of Pb involved reaction with Fe/Mn oxides and organic matter. On the other hand, the removal of Pb is attributed significantly to the retention by hydrous oxides of Fe/Mn phases and the organic phase. Compared to the other heavy metals, the removal of Pb from the exchangeable fraction is negligible in almost all spent media. The observed results may be either due to strong sorption of Pb or slow release of sorbed Pb. Hydrous oxides of Mn and Fe that are well known sinks in the surface environment for heavy metals are extracted together. Retention of metals by these secondary oxides, present as coatings on mineral surfaces or as fine discrete particles, can occur by any or a combination of the following mechanisms: co precipitation, adsorption, surface complex formation, ion exchange, and penetration of the lattice (Hall et al. 1996). The precipitation of Pb as $PbSO_4$ cannot be overlooked too as a possible removal mechanism of Pb from the simulated polluted water. Sulphate anions are released from $MgSO_4.7H_2O$, which is one of the constituents used to prepare the simulated polluted water.

Figure 5.26 shows how Pb, Cu, and Zn desorbed from the filter media with strong acid (acid exchangeable) distributed in different depth sections of the spent media after leaching them with 40 ml of (0.11 mol/l acetic acid). The depth - profiles clearly show that the exchangeable fractions of Pb, Cu, and Zn have been distributed in similar fashion in the columns where Zn showing the highest mobility from the spent filter media.

Significant amounts of Ni and Zn (18%) are also retained in the reducible fraction of spent media sampled from column filled with MnR sediment, which might involve precipitation of metal hydroxides with Mn/Fe oxihydroxides. In spent media taken from columns filled with KVA volcanic ashes and QTZ sand 20% of these metals are also retained in Fe/ Mn oxide phases while in columns filled with DQ scoria and EBF laterite removals of Ni, Cu, Zn, and As are mainly associated with the exchangeable fraction (about 80%) while the remaining 15 - 20 % are held in the Fe/Mn oxide phases. The association of Pb in the Fe/Mn oxide phases reach up to a maximum of 35%, represents the highest proportion compared to the other heavy metals. Hydrous oxides of Mn and Fe are released together, which are the well-known 'sinks' in the surface environment for heavy metals (Stone et al. 1996).

The association of heavy metal ions with a ferric or aluminium hydroxide precipitate is controlled by adsorption or coprecipitation. Adsorption implies the two-dimensional accumulation of a metal ion at the interface between a solid surface and the aqueous phase. Coprecipitation implies the simultaneous removal of a metal ion during the formation of primary metal precipitates (Lee et al. 2002). Due to the potential for adsorption, coprecipitation and the ion exchange characteristics of the colloidal hydroxide precipitates, even clear differences in pH for hydroxide formation are no guarantee avoiding mixed precipitates when metal mixtures are in solution (Brooks 1991).

Sulfides such as Na_2S, H_2S, NaHS, or FeS can be used to precipitate metals and have been applied frequently for metal removal from waste effluents to achieve emission standard limitations (Brooks 1991, Peters et al. 1988). The lower solubility of metal sulfides in the acid region below pH value of 7 permits reduction of metal solubility to values that are orders of magnitude lower than are attainable by hydroxide precipitation (Brooks 1991).

Sequential extraction yields for Cu, Zn, Ni and As are negligible in the oxidizable/ organic fraction of the spent media. The organic fraction of metals released in the oxidizable step is not considered very mobile or available since it is thought to be associated with stable high molecular weight humic substances that release small amounts of metals in a slow manner (Miragaya et al. 1994). This fraction is one of the smallest or even negligible in the entire spent media extracted. However, this is an important fraction, especially in sludges where considerable amount of domestic wastes are released. In this case the organic fraction even dominates trace metal distribution. Compared to the other heavy metals higher proportion of Pb is associated with the oxidizable phase (about 10% are associated to the organic phases) (Fig. 5.25). Peptone, meat extract and urea are the main composition of the synthetic wastewater, which are mainly of organic composition that might interfere with the adsorptive filtration process by interacting with either the metals or the surface of the media. Figure 5.25 shows the column performance in removing Pb from solution in the presence of these organic compounds as compared to the other heavy metals.

Minor amounts of Zn and Pb (<1%) have been released into the effluents after the spent media were leached with tap water showing that disposal of the used filter media do not pose environmental pollution such as of soil, groundwater and surface water under natural hydrochemical conditions. Since these metals bound to the water -soluble fraction of the spent media are very low, mobilization by rainwater or surface runoff is insignificant. Trace elements extracted by H_2O are relatively labile and thus may be potentially bioavailable (He et al. 1995).

The water-soluble fractions of Cu and Zn are 1% of the corresponding Cu and Zn fractions in all column sludges when expressed as absolute amounts (mg/kg). Hall et al. (1996) indicated that the water-soluble fraction of a soil or sediment is the first to be released and is usually negligible, except in areas where evaporites are present. Results of the single water extraction procedure confirmed that the removal of most of heavy metals from the mobile fraction is confined to the first 5 - 10 cm in the columns (Fig. 5.24). The most likely explanation is that the high porosity facilitated thc trapping of SPM within this depth profile where the availability of adsorption sites would be enhanced.

The fractionation and binding behaviours of all heavy metals considered in this study are summarized in figure. 6.5. From the diagrams it is possible to observe that the acid exchangeable fractions of Cu, Zn, Ni, As and Pb seem to distribute evenly in all spent media except in MnR sediment where the fractionation of Pb is the lowest. The reducible fraction of Pb also show even distribution in all used filter media except in PYC pyroclastics where Pb fractionation is higher in the latter. Results of leaching experiments also indicate that, As is the most mobile element when the filter media is treated with tap water, an indication that As is weakly retained on the surfaces of the filter media (Fig 6.5).

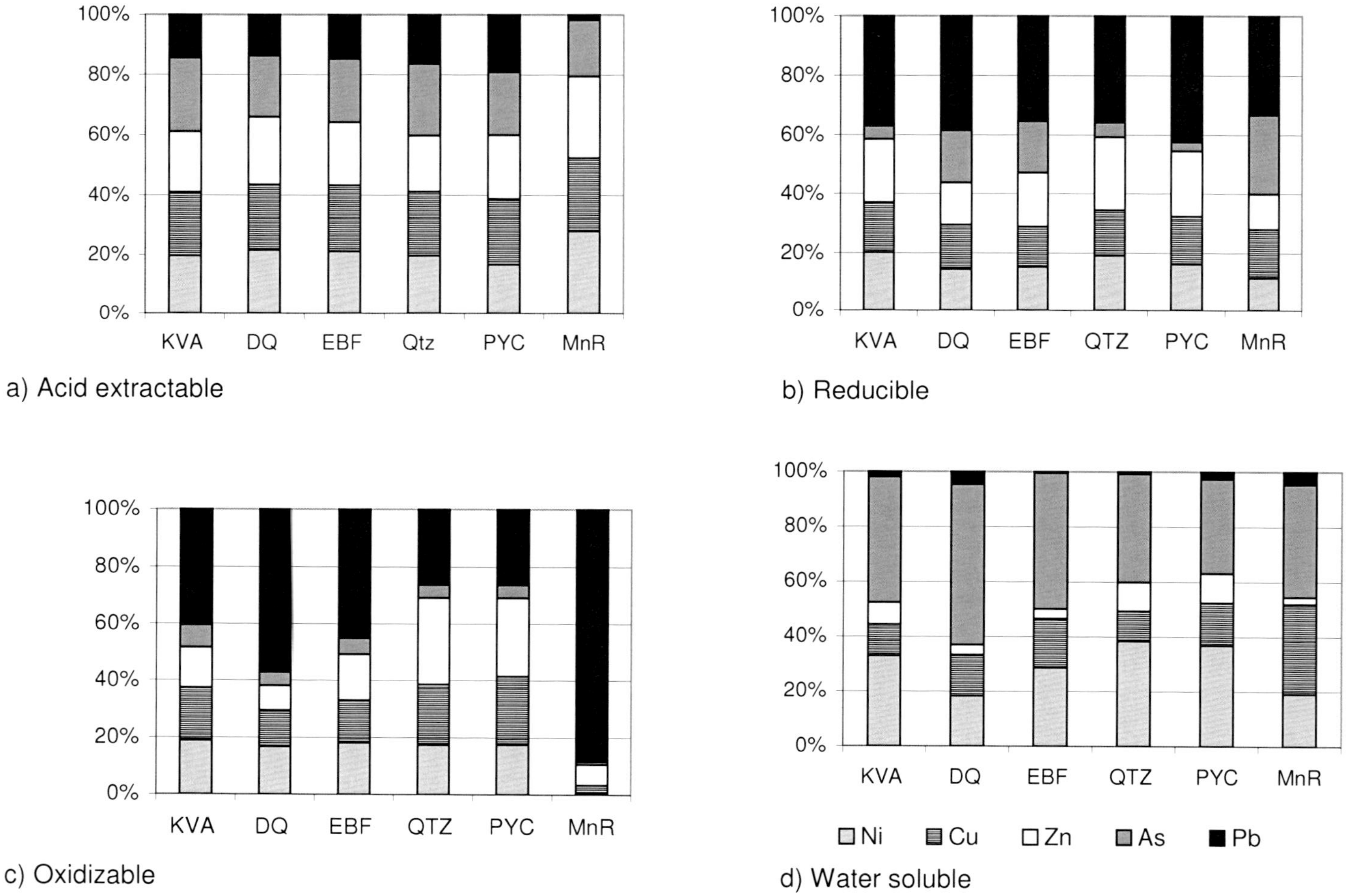

Fig. 6.5. Comparison of the fractionation of Ni, Cu, Zn, As, and Pb in the spent media (0 - 10 cm).

Many researchers have shown increased Cd and/or Pb sorption in tropical soils and/or in pure oxidic mineral systems (e.g. goethite) with increasing pH due mainly to increased negative surface charge, (Naidu et al. 1994). A gradual increase of the pH of effluents from filter columns receiving simulated polluted water (SPW) with increasing numbers of pore volume displacements was observed by Fuller et al. (1976). In the present study the frequent flow of SPW through the columns might have temporarily caused anaerobic conditions, which are believed to be responsible for the pH increase. Since reduction process removes hydrogen ions from the solution, pH of the effluents increases.

The rise in pH of influents passing through the column media is another factor that contributed to the formation of different hydroxo species based on the hydrolysis constants of different metal ions as defined in

$$M^{2+} + nH_2O \rightleftharpoons M(OH)_n^{\ 2-n} + nH^+$$

and taking only primary metal species expected to be formed. Schwertmann and Taylor (1989) postulated that pH is the main force governing the adsorption of metal cations, and the fact that pH of maximum increase in adsorption is found to be linearly related to the first hydrolysis constant of the metal $K_1 = (MOH^+)/(M^{2+})\cdot(OH^-)$ indicates that the hydrolyzed species (MOH^+) is preferentially adsorbed over the unhydrolyzed one (M^{2+}). At very high pH values, oxo ions (O_2^-) may occur on the surface of an oxide or silicate in contact with an aqueous solution. Hsu (1989) also suggested that the exchange retention should be viewed as the competition between M^{2+} and H^+ for the surface O on the basis of the cation relative affinity for this surface O. Since the pH conditions at which the column experiments have been carried out are in the basic range it might be possible to state that the removal of Pb, Zn, Ni, and Cu is mostly due to retention on minerals such as SiO_2, Fe and Mn oxides. However, there may also be a possibility of adsorption of smaller fractions of heavy metals on mineral surfaces. The other important factor that might be the reason to higher retention of heavy metals with increased pH, is the pH_{PZC} of the filter media. The wide range of pH_{PZC} of the important minerals in the filter media resulted in differences in charge characteristics that affect the exchange reactions and removals from solid solutions (Table 6.6). At any pH below pH_{PZC} the surface of metal oxides/oxyhydroxides is positively charged and at pH above pH_{PZC} the surface is negative (Parida et al. 1996). When the solution pH exceeds PZC, the metal species are more easily attracted to the negatively charged surfaces of the filter media, favoring accumulation of metal species on the surface and thus facilitating retention.

Table 6.6. pH_{ZPC} of minerals present in the filter media (modified after Parks 1965).

Material	pH_{ZPC}
α - Al_2O_3	9.1
α - $Al(OH)_3$	5.0
γ Fe_2O_3	6.7
MnO_2	2 - 4.5
SiO_2	2.0
Kaolinite	4.6
Montmorillonite	2.5
Albite	2.0

Points of zero charge of silicate and oxide surfaces have been evaluated in several studies (Parks 1965, Sverjensky 1994) and range from 2 - 3 (SiO_2) to ≈ 12 (MgO). Lead, and Cu are the heavy-metal cations most strongly retained by surface of the filter media whereas Ni, and As are the relatively least strongly retained in this competitive situation. The competition for sorption surfaces among metal ions is also another controlling factor that determines the sorption of metals. According to Frost and Griffin (1978) precipitation contributes significantly to the removal of Cu from leachate at pH values above 5, while for Zn, precipitation affects removal only once the pH exceeds 7. In this present study precipitation and adsorption of Cu by filter media has also direct relation with the rising of pH. Adsorption and precipitation could be possible mechanisms responsible for the removal of Cu, Ni, and Zn in columns packed with volcanic materials. The main sorbing surfaces in the volcanic ashes may be associated to the oxides of Al, Fe and clay minerals. The porous and vesicular nature of scoria might also contribute to the available sorbing surfaces in retaining some of the heavy metals. In DQ scoria hematite and kaolinite played a significant role in metal removal while in EBF laterite, kaolinite and montmorillonite contributed to the removal of metals.

Arsenate and arsenite adsorb to surfaces of a variety of aquifer materials, including Fe oxides, Al oxides, and clay minerals. Adsorption of both As(III) and As(V) from groundwater on a low cost ferruginous Mn ore in the pH range of 2 - 8 were shown (Chakravarty 2002). Some studies have shown that the adsorption and oxidation of As (III) can occur on the surface of Fe hydroxides, e.g., goethite, a-FeOOH (Sun and Doner 1996, 1998), ferrihydrite (Raven et al. 1998), and Mn oxides, e.g., birnessite, α-MnO_2 (Scott and Morgan 1995). Similarly adsorption of As to Fe oxide surfaces tends to decrease as pH increases, at least between the ranges from pH 6 to pH 9 (Dzombak and Morel 1990). The pH-dependence of AsO^{3-}_4 adsorption to Fe oxide surfaces appears to be related to the change in Fe oxide net surface charge from positive to negative as pH increases above the zero-point-of-charge of about 7.7 for goethite (crystalline Fe oxide) (Stumm and Morgan 1996) or 8.0 for ferrihydrite (amorphous Fe oxide) (Dzombak and Morel 1990). Where pH values are above 8, the negative net surface charge of Fe oxide can repel negatively charged ions such as arsenate. In all columns the effluents pH increased from an initial solution pH of 7 to 8.5 ± 0.1, which might contribute to the poor retention of As by Fe-oxide and/or clay surfaces of the DQ scoria and EBF laterite.

6.4.3 Release of Heavy Metals and Arsenic

Although DQ scoria and EBF laterite columns removed the spiked metals from synthetic wastewater, the release of Fe and Mn from these columns should not be overlooked. These media are characterized by high contents of basic minerals such as clinopyroxene, hematite, and clay minerals that could be sources of Fe and Mn minerals. Release from KVA volcanic ash is also significant. In columns filled with KVA volcanic ash, the highest release of Fe and Mn reveals some correlations with the low dissolved oxygen concentrations. Compared to the DQ scoria and EBF laterite columns the desorption of Fe and Mn from columns filled with MnR sediment, PYC pyroclastics, and QTZ sand are not significant. The lowest desorbed Mn concentrations are measured in column packed with MnR sediment despite the fact that this sediment has high content of Mn oxide (birnessite). This is an indication that Mn is precipitated with the other heavy metals under the prevailing basic pH condition (pH = 8 - 8.5).

When Mn (II) is present in aerobic environments, a complex series of oxidation/precipitation and adsorption reactions occurs, which eventually renders the Mn biologically unavailable as insoluble Mn dioxide. However, the kinetics of Mn (II) oxidation is slow in waters with pH below 8.5 (Zaw & Chiswell 1999). The time required for the oxidation and precipitation of Mn ranges from days in natural waters to years in synthetic waters (Stokes et al. 1988). However, oxidation rates of Mn increase with increasing pH or the presence of catalytic surfaces such as Mn dioxide (Huntsman & Sunda 1980). In a stream receiving Mn-rich inflows caused by acid mine drainage, there was rapid oxidation and precipitation of Mn oxides (Scott et al. 2002). The sequence of reactions involving the oxidation of Mn (II) and subsequent precipitation as Mn dioxide includes simultaneous occurrence of several Mn forms (i.e., dissolved Mn (II), hydrous oxides of Mn (III), Mn (II) adsorbed to particulates, and Mn (II)-ligand complexes), with individual concentrations dependent on factors that include pH, inorganic C, organic C, SO_4^{2-}, Cl^-, temperature, and time (Stokes et al. 1988). In groundwater with low oxygen levels, Mn (IV) can be reduced both chemically and bacterially to the Mn (II) oxidation state (Jaudon et al. 1989).

The lower dissolved oxygen contents measured between day 44 and 46 (< 3 mg/l) are indicators of the reducing condition in all columns that in turn resulted in the release of Fe and Mn (Fig. 5.20). Manganese and Fe oxides are among the main electron acceptors that allow bacterial activity in soils under anaerobic conditions, and Mn and Fe solubilization is a direct result of anaerobic bacterial activity (Quantin et al. 2001). Solubilization is greatly stimulated by organic matter as observed in this experiments where Mn and Fe release reach quite high levels. The reduction of Mn and Fe also resulted in the desorption of As associated to oxi/hydroxides of DQ scoria and EBF laterite where Fe and Mn contents are substantial. Mn and Fe oxides, which are favorable for the sorption of Cu and specially Zn (Hickey and Kittrick 1984, Ramos et al. 1994) can, under anaerobic conditions accept electrons. Their reduced forms are mobile and can be, at least partially removed with the effluent. The rise in pH in the columns could also play a role in the desorption of As from columns filled with DQ scoria and EBF laterite after day 44. Under aerobic conditions, oxygen acts as electron acceptor for the respiratory process. In the absence of free oxygen, however, a number of other substances such as NO_3^-, Mn and Fe compounds, H ions, and SO_4^{2-} can accept electrons and take part in reduction reaction.

Release of Mn and Fe from the columns in turn favored the simultaneous release of As that has been already precipitated with the oxides of Fe and Mn at higher pH values. According to Fuller and Davis (1989), Dzombak and Morel (1990) the desorption of AsO_4^{3-} from Fe oxide surfaces becomes favored as pH values become alkaline.

6.5 Applicability

In considering the removal of Zn and Cu from the mixed simulated polluted water passed through the column packs (Fig. 5.19), the four filter media show similar relative retention behavior, e.g. retention slightly decreases in the following order for Zn and Cu: MnR sediment > EBF laterite > DQ scoria > KVA volcanic ash ≥ QTZ sand. Higher SPM concentrations in Akaki surface water facilitate the association of metals with SPM, which later could be purified by media filtration. Few available sorption sites in the volcanic materials make them poor sorbents to retain As. The higher desorption of As from KVA volcanic ashes QTZ sand and PYC pyroclastics make them unsuitable for As removal. On the other hand, the desorption of Mn and Fe from the volcanic materials are less significant as compared to columns filled with DQ scoria and EBF laterite. Therefore, these media can directly be used in the treatment of heavy metal laden water in the Akaki area.

Surface water and groundwater of low As content could be purified by using filter media constituted by KVA volcanic ashes where the desorption of Fe and Mn is not substantial. The application of DQ scoria and EBF laterite in water purification is also recommended provided that the desorption of Fe and Mn in to the effluents can be minimized (e.g. aeration of the effluents during purification). Regarding the filter media collected from Indonesia, MnR sediment can be used to treat heavy metals including As polluted water. On the other hand, the removal capacity of PYC pyroclastic for As is very poor. The performance in removing Ni is also variable and decreasing with time. Application of PYC pyroclastic as filter media to treat As and Ni polluted water is not recommended.

Overall, the project adequately demonstrated the potential applicability of volcanic materials and their weathering products as filter media, and it appears that the filter media are appropriate for testing at a larger scale.

7 Conclusion

In the Akaki Sub-Basin the weathering products of volcanic rocks as well as the circulation of thermal water that is found at the center of Addis Ababa significantly affect the concentrations of chemical compounds in surface water and groundwater. The other factors contributing to the change in the physico-chemical characteristics of the water are the anthropogenic inputs of domestic and industrial wastes, which enrich certain anions and cations such as Na^{+}, Cl^{-}, NO_3^{-}, SO_4^{2-} and some heavy metals.

Water parameter data showed that though most groundwater samples in the Akaki Sub-Basin show fairly normal values of dissolved solutes, surface water samples revealed higher amounts due to the influence of anthropogenic pollution. Dissolved oxygen as low as 0.2 mg/l is also measured in surface water samples indicating that these waters are highly polluted. It has also been observed that the suspended load was much higher in the lake and the two river systems, which resulted from surface runoff upstream of the study area.

The type of water in the Akaki Sub-Basin is alkaline earth fresh water with higher contents of alkalis, prevailing hydrogen carbonate. Anionic measurements showed that nitrate is found in concentrations exceeding the U.S. Environmental Protection Agency (US-EPA) maximum contaminant limit (MCL). In 50% of all groundwater water samples collected, the concentrations of nitrate are higher than the US-EPA MCL (10 mg/l). Comparison of the concentrations of major cations and anions of the Akaki water in relation to that of the estimated mean composition of river water of the world showed that concentrations of Na and K in surface waters of Akaki are ten times higher than those of the mean composition of the World Rivers. Magnesium and Ca concentrations are also two to three times higher than the mean composition of world river waters indicating that the Akaki Sub-Basin is enriched in major cations derived from both geogenic and anthropogenic sources.

The concentrations of heavy metals in the dissolved phase are generally lower in most surface waters than in groundwater. In the dissolved phase, high concentrations are measured for Mn and Fe. The concentrations of Fe and Mn in the dissolved phase are well above the Secondary Standards that are set by the US-EPA (0.3 mg/l and 0.05 mg/l for Fe and Mn respectively). In general, very low concentrations of dissolved as well as heavy metals bound to SPM are measured in the Akaki Well fields.

However, comparison among suspended particulate matter bound and dissolved phase concentrations of heavy metals show that very high concentrations of heavy metals are detected on the particulate matters. The concentrations of Cu, Zn, Ni, As and Pb bound to SPM in groundwater samples are higher than that of the surface water.

In surface water samples, heavy metals Cu, Zn, and Pb concentrations on the suspended particulate matter are higher than the values stated in the LAWA classes for moderately contaminated SPM (Class II). On the other hand, in almost all water samples concentrations of Cu, Zn, and Pb bound to SPM are higher than the threshold value (class I) that shows no contamination. The concentrations of Cu, Zn, and Pb bound to SPM in groundwater water samples are well above the limit indicated for Class III (strongly contaminated) in the LAWA classes.

The distribution coefficients (K_d) of heavy metals calculated for the two phase system (water-solid) reveal that the particulate matters bound Fe and Mn concentrations are so high that the values can not be directly compared to the Cu, Zn, Ni, Pb and As. Compared Cu and Zn bound to SPM As and Pb bound to SPM showed relatively lower concentrations. The strong association of heavy metals with the particulate matter in surface water and groundwater of the Akaki Sub-basin leads to investigate an appropriate purification technique where both dissolved and heavy metals bound to SPM can be removed from the polluted water.

In water samples collected in 2003 suspended particulate matter (SPM) and several trace elements are detected more commonly and in higher concentrations in Akaki rivers and Aba Samuel Lake than at sites that are located upstream of the catchment area and the Akaki well field. The need to purify these water sources in the downstream side of the basin is critical.

From twelve materials of different chemical and mineralogical composition, three natural materials and their weathering products derived from Ethiopia were selected for in depth study of heavy metal retention capacities based on their proximity to the target area, natural background heavy metals concentration (below the permissible MCL levels set by US EPA), appropriate grain size. The results of XRD, XRF, and sequential extractions of these selected filter media showed that KVA volcanic ash, DQ scoria, MnR sediment and EBF laterite are potential filter materials in removing heavy metals from polluted surface and groundwater. Sequential extraction of the filter media indicated that the mobile fraction of heavy metals associated with DQ scoria, KVA volcanic ash, and EBF laterite are below the detection limit (<0.1 mg/l). These materials are found to be appropriate for laboratory column experiments in removing heavy metals from simulated polluted water.

Results of the column experiments revealed that the quality of effluents was enhanced in terms of suspended particulate matter concentrations, concentrations of dissolved and heavy metals bound to SPM, and some water parameters. The removals of SPM and thereby heavy metals bound to SPM are effective for all filter media tested.

The MnR sediment is found to be the best performing media in removing almost all of the metals from the synthetic solutions. However, considering the filter media collected from Ethiopia the EBF laterite and DQ scoria are found to be better performing media in removing almost all of the metals from the synthetic solutions except As. However, the release of Fe and Mn from these columns is significant and should be considered before applying the media for water treatment. Simulated polluted water of low As content could also be purified using filter media constituted with KVA volcanic ashes and QTZ sand where the desorption of Fe and Mn is not substantial.

Sequential extraction of the spent media indicated that more than 70% of Cu, Zn, and Ni are bound to the acid exchangeable fraction while Pb showed variable fractionation among the three fractions defined in BCR sequential extraction scheme. From the BCR extraction procedure it is possible to observe that the metal removal mechanisms mainly involved weak retentions including weakly adsorbed metals retained on the surface of the filter media by relatively weak electrostatic interaction and metals that can be released by ion-exchange processes. Results of single step water extraction of the used media showed that the heavy metals of interest are less likely adsorbed in the water-soluble fraction. From the results obtained it can be concluded that the disposal of the spent media in agricultural fields or the use as construction material do not pose environmental pollution. Pollution of surface water and groundwater by heavy metals incorporated in the sludge is minimal under natural hydrochemical conditions.

Polluted surface water and groundwater of the Akaki Sub-Basin with low As content could be purified by using a mixture of sand and KVA volcanic ash. The application of DQ scoria and EBF laterite in water purification is also recommended provided that the desorption of Fe and Mn in to the effluents can be minimized (e.g. aeration of the effluents during purification). MnR sediment can be used to treat heavy metals including As polluted surface water and groundwater effectively.

The overall removal capacities of the filter media for heavy metals proved to be promising. However, the heavy metal removal capacities of the filter media on a pilot scale filter system could be much more complex than what was performed under laboratory condition. The removal of heavy metals using local materials might show different results if the Akaki raw water were used as the influent solution during the experiment. For a better and applicable results heavy metal removal column experiments using the polluted water of Akaki is recommend.

8 References

AAWSA, AAU, (2003): Groundwater vulnerability mapping of the Addis Ababa water supply aquifers, Ethiopia. Report to UNEP/UNESCO/UN-HABITAT/ECA, 80p.

AAWSA, BCEOM Seureca-Space in Association with Tropics Consulting Engineers. (2000): Akaki groundwater model Report. Report V. I-IV.

AAWSA-SEURICA, (1991): Addis Ababa Water Supply Project Stage III. Feasibility study and preliminary design. Ground Water Resources. V.IV. Addis Ababa.

ADRIANO, D.C. (2001): Trace elements in terrestrial environments: Biogeochemistry, Bioavailability, and Risks of Metals, 2nd edition, Springer-Verlag, New York, 867p.

AHMAD, S.; KITCHIN, K.T. & CULLEN, W.R. (2000): Arsenic species that cause release of iron from ferritin and generation of activated oxygen, Arch. Biochem. Biophys., 382: 195-202.

AHNSTROM, Z.S.; PARKER, D.R. (1999): Development and assessment of a sequential extraction procedure for the fractionation of soil cadmium. Soil Sci. Soc. Am. J., 63: 1650-1658.

ALABASTER, J.S. and LLOYD. R. (1982): Water quality criteria for freshwater fish. Butterworth Scientific, London, 361p.

ALEMAYEHU, T. (2001): The impact of uncontrolled waste disposal on surface water quality in Addis Ababa, Ethiopia. SINET: Ethiop. J. Sci., 24: 93-104.

ALEMAYEHU, T. (2004): Assessment of pollution status and groundwater vulnerability mapping of the Addis Ababa aquifers, Ethiopia, 12p.

ALLEN, J.P and TORRES, I.G. (1991): Physical separation techniques for contaminated sediment. In: Li, N.N. (Eds.): Recent Developments in Separation Science. CRC Press, West Palm Beach, FL, Vol V, 81p.

ALLISON, J.D.; ALLISON, T.L. (2005): Partition coefficients for metals in surface water, soil, and waste. USEPA, Office of Research and Development Washington, DC 20460, 93p.

ALLOWAY, B.J. (1990): Sorption of trace metals by humic materials in soil. In: ALLOWAY, B.J. (Eds.): Heavy metals in soils. Blackie, Glasgow, 339p.

ALLOWAY, B.J. (1990): Heavy metals in soils. Blackie, London, 339p.

ALLOWAY, B.J.; AYRES, D.C. (1997): Chemical principles of environmental pollution. CRC Press, 56p.

ANGELONE, M.; BINI, C. (1992): Trace elements concentrations in soils and plants of Western Europe. In: ADRIANO, D.C. (Eds.): Biogeochemistry of trace metals. Lewis Publishers, 19-60.

BAES, C.F., Jr. & MESMER, R.E. (1976): The hydrolysis of cations. John Wiley and Sons, New York, 189p

BAEYENS, B. & BRADBURY, M.H. (1995): A quantitative mechanistic description of Ni, Zn and Ca sorption on Na-montmorillonite. Report PSI Ber., Paul Scherrer Institut, Villigen, Switzerland, 95-11.

BAILEY, S.E.; OLIN, T.J.; BRICKA, R.M.; ADRIAN, D.D. (1999): A review of potentially low costs sorbents for heavy metals. Water Res., 33: 2469-79.

BAKER, J.P. (1991a): Biological effects of changes in surface water acid-base chemistry. In: IRVING, P.M., (Ed.): Acid deposition: state of science and technology. Report 13. Summary report of the National Acid Precipitation Assessment Program, Washington, D.C., 119-123

BAKER, L.A. (1991b): Current status of surface water acid-base chemistry. In: IRVING, P.M., (Ed.): Acid deposition: state of science and technology. Report 9. Summary report of the National Acid Precipitation Assessment Program, Washington, D.C., 89-101.

BAKER, D.E.; SENFT, J.P. (1995): Copper. In: ALLOWAY, B.J. (Ed.): Heavy metal in soils. 2nd ed. Blackie Academic and Professional, 11-37.

BASTA, N.T.; TABATABAI, M.A. (1992): Effect of cropping systems on adsorption of metals by soils: III. Competitive adsorption. Soil. Sci., 153: 331-337.

BELL, F.G. (1998): Environmental Geology. Principles and Practice. Blackwell Publishing, 263pp.

BENJAMIN, M.M. & LECKIE, J.O. (1980): Adsorption of metals at oxide interfaces: effects on the concentration of adsorbate and competing metals. In: BAKER, R.A. (Ed.): Contaminants and sediments, Volume 2, Ann Arbor Science, Ann Arbor, Michigan, 305-332.

BENOIT, G.; OKTAY-MARSHALL, S.D.; CANTU, A.; HOOD, E.; COLEMAN, C.; CORAPCIOGLU, M. & SANTSCHI, P. (1994): Partitioning of Cu, Pb, Ag, Zn,

Fe, Al and Mn between filter retained particles, colloids and solution in 6 Texas estuaries. Mar. Chem., 45: 307-336.

BERMANEC, V.; SIJARIC, G.; KNIEWALD, G. (2000): Gaspéite and associated Ni-rich minerals from veins in altered ultrabasic rocks from Duboštica, Bosnia and Herzegovina. The Can. Mineral., 38: 1371-1376.

BERNARD, A.M.; VYSKOCIL, A.; KRIZ, J.; KODL, M. & LAUWERYS, R. (1995): Renal effects of children living in the vicinity of a lead smelter. Environ. Res., 68: 91-95.

BERNER, R.A. (1971): Worldwide sulfur pollution of rivers: J. Geophys. Res., 76: 6597-6600.

BERNER, E.K.; BERNER, R.A. (1987): The Global water cycle: Geochemistry and Environment, Prentice-Hall, Inc. Englewood Cliffs, N.J., 397p.

BETHKE, C.M. (1996): Geochemical reaction modeling. Concepts and Applications. New York, NY: Oxford University Press, 416p

BHATTACHARYA, A.K.; VENKOBACHAR, C. (1984): Removal of cadmium (II) by low cost adsorbents. J. Environ. Eng., 110: 110-122.

BIBBY, R.L.; WEBSTER - BROWN, J.G. (2005): Characterization of urban catchment suspended particulate matter (Auckland region, New Zealand): a comparison with non-urban SPM. Sci. Tot. Environ., 343: 177-197.

BLEEKER, P. & SAGEMAN, R. (1990): Surface charge characteristics and clay mineralogy of some variable charge soils in Papua New Guinea. Aust. J. Soil. Res., 28(6): 901-917.

BODEK, I.; LYMAN, W.J.; REEHL, W.F. & ROSENBLATT, D.H. (1988): Environmental inorganic chemistry: Properties, processes and Estimation Methods, Pergamon Press, Elmsford, NY, 671p.

BOULLEMANT, A.; VIGNEAULT, B.; FORTIN, C.; CAMPBELL, P.G.C. (2004): Uptake of neutral metal complexes by a green alga: influence of pH and humic substances. Aust. J. Chem., 57: 931-936.

BROCK, S.L.; DUAN, N.G.; TIAN, Z.R.; GIRALDO, O.; ZHOU, H. & SUIB, S.L. (1998): A review of porous manganese oxide materials. Chem. Mater., 10(10): 2619-2628.

BROOKINGS, D.G., (1988): Lead. In: Eh-pH diagrams for geochemistry. Heidelberg: Springer-Verlag, 42-43.

BROOKS, C.S. (1991): Metal recovery from industrial waste. Lewis Publishers, Inc. Chelsea, Michigan, USA, 215p.

BROWN, R.B. (2003): Soil texture. Department of Agriculture, Cooperative Extension Service, University of Florida, IFAS, Florida (http://edis.ifas.ufl.edu).

BROWN, P.A; JEFCOAT, I.A.; PARRISH, D.; GILL, S.; GRAHAM, E. (2000a): Evaluation of the adsorptive capacity of peanut hull pellets for heavy metals in solution. Adv. Environ. Res., 4: 19-29.

BROWN, P.A.; GILL, S.A.; ALLEN, S.J. (2000b): Metal removal from wastewater using peat. Water Res., 16: 3907-3916.

BUOL, S.W.; HOLE, F.D; MCCRACKEN, R.J.; SOUTHARD, R.J. (1997): Andisols: Soils with andic soil properties. In: Soil Genesis and Classification (4th ed.), Iowa State University Press, 249-257.

CALLAHAN, M.A.; EHRETH, D.J. & LEVINS, P.L. (1979): Municipal sludge management. Proc. 8th Natn. Conf., Miami, U.S.A., 55.

CAMPY, M. ; MEYBECK, M. (1995) : Les sédiments lacustres. In: POURRIOT, R. & MEYBECK, M. (Eds.): Limnologie Générale. Masson, Paris, 185-226.

CHAKRAVARTY, S.; DUREJA, V.; BHATTACHARYYA, G.; MAITY, S.; BHATTACHARJEE, S. (2002): Removal of arsenic from ground water using low cost ferruginous manganese ore. Water Res., 36: 625-632.

CHAPMAN D. (1996): Water quality assessments - A guide to use of biota, sediments and water in environmental monitoring, 2nd Edition, 134.

CHASHSCHIN, V.P.; ARTUNIA, G.P. & NORSETH, T. (1994): Congenital defects, abortion and other health effects in nickel refinery workers. Sci. Total Environ., 148: 287-291.

CHEN, Y.C.; LIN-SHIAU, S.Y. & LIN, J.K. (1998): Involvement of reactive oxygen species and caspase 3 activation in arsenite-induced apoptosis, J. Cell. Physiol., 177: 324-333.

CHING, S. & SUIB, S.L. (1997): Synthetic routes to microporous manganese oxides. Comments on Inorganic Chemistry 19(5): 263-282.

CHISHOLM - BRAUSE, C.J.; CONRADSON, S.D.; BUSCHER, C.T.; ELLER, P.G.; MORRIS, D.E. (1994): Speciation of uranyl sorbed at multiple binding sites on montmorillonite. Geochim. Cosmochim. Acta, 58: 3625-3631.

CHILLRUD, S.N.; BOPP, R.F.; SIMPSON, H.J.; ROSS, J.M.; SHUSTER, E.L.; CHAKY, D.A.; WALSH, D.C.; CHOY, C.C.; TROLLEY, L-R. & YARME, A. (1999): Twentieth century atmospheric metal fluxes into Central Park Lake, New York City. Environ. Sci. Technol., 33: 657-662.

CLEVENGER, T.E. (1990): Use of sequential extraction to evaluate the heavy metals in mining wastes. Water Air Soil Pollut., 50: 241-254.

CLEVENGER, T.E. & MULLINS, W. (1982): The toxic extraction procedure for hazardous waste. In: Trace substances in environmental health XVI. Univ. of Missouri, Columbia, MO., 77-82.

COETZEE, PP. (1993): Determination and speciation of heavy metals in sediments of the Hartbeespoort Dam by sequential chemical extraction. Water SA., 19(44): 291-300.

CORNELL, R.M.; SCHWERTMANN, U. (1996): The iron oxides. VCH, Weinheim, 573p.

DANIEL, G. (1977): Aspects of climate and water budget in Ethiopia, Technical monograph, Addis Ababa University Press, Addis Ababa, 393p.

DAVIDSON, C.M.; FERREIRA, P.C.S. & URE, A.M. (1999): Some sources of variability in application of the three-stage sequential extraction procedure recommended by BCR to industrially contaminated soil. Fresenius J. Anal. Chem., 363: 446-451.

DAVIDSON, C.M.; GOMEZ, A.; LUCK, D.; BACON, J.; YLI-HALLA, M.; MUNTAU, H. & QUEVAUVILLER, PH. (2000): Application of a modified BCR sequential extraction (three-step) procedure for the determination of extractable trace metal contents in a sewage sludge amended soil reference material (CRM 483), complemented by a three-year stability study of acetic acid. J. Environ. Monit., 2: 228-233.

DAVIS, J.A.; KENT, D.B. (1990): Surface complexation modeling in aqueous geochemistry. In: HOCHELLA, M.F. JR.; WHITE, A.F., (Eds.): Mineral-water interface geochemistry. Reviews in mineralogy. Washington, D.C. The Mineral. Soc. Am., 23: 177-260.

DEANS, J.R.; DIXON, B.G. (1992): Uptake of Pb^{2+} and Cu^{2+} by novel biopolymers. Water Res., 26(4): 469-472.

DEPLEDGE, M.H.; FOSSI, M.C. (1994): The role of biomarkers in environmental assessment (2). Invert. Ecotoxicol., 3: 161-172.

DEUTSCHES INSTITUT FÜR NORMUNG. (1984): German standard procedure for water and sludge analysis, sludge and sediments (group S). Determination of water leachability. DIN 38414-S4, Benthe-Vertrieb, Berlin.

DZOMBAK, D.A. & MOREL, F.M.M. (1990): Surface Complexation Modeling: Hydrous Ferric Oxide, John Wiley & Sons, New York, 393p.

ECHEVERRÍA, J.C.; MORERA, M.T.; MAZKIARÁN, C.; GARRIDO, J.J. (1998): Competitive sorption of heavy metal by soils. Isotherms and fractional factorial experiments. Environ. Pollut., 10: 275-284.

EISMA, D. (1993): Suspended matter in the aquatic environment. Springer-Verlag, Berlin, Germany, 315p.

ELLIOTT, H.A.; LIBERATI, M.R.; HUANG, C.P. (1986): Competitive adsorption of heavy metals by soils. J. Environ. Qual., 15: 214-219.

ELZINGA, E.J. & SPARKS, D.L. (1999): Nickel sorption mechanism in a pyrophyllite-montmorillonite mixture. J. Coll. Inter. Sci., 213: 506-512.

ELZINGA, E.J. & SPARKS, D.L. (2001): Reaction condition effects on nickel sorption mechanisms in Illite-Water suspensions. Soil Sci. Soc. America J., 65: 94-101.

ESTÈBE, A.; MOUCHEL, J.M. & THÉVENOT, D.R. (1998): Urban runoff impacts on particulate metal concentrations in River Seine. Wat. Air Soil Pollut., 108: 83-105.

ETHIOPIAN ENVIRONMENTAL PROTECTION AUTHORITY (EEPA) (1997): Preliminary survey of pollutant load on Great Akaki, Little Akaki and Kebena Rivers. Addis Ababa.

ETHIOPIAN CENTRAL STATISTICAL AUTHORITY (1999): The 1994 Population and Housing Census of Addis Ababa. Analytical Report. V.II. Addis Ababa.

ETHIOPIAN MAPPING AGENCY, GEOGRAPHY DIVISION (1981): National Atlas of Ethiopia. 1:4000000.

EVANKO, C.R. & DZOMBAK, D.A. (1997): Remediation of metals-contaminated soils and groundwater. Groundwater Remediation Technologies Analysis Center. Technology Evaluation Report. Pittsburgh, PA, 61p.

EVANS, R.C. (1966): An introduction to crystal chemistry. Cambridge University Press, London, 410p.

EVANS, L.J. (1989): Chemistry of metal retention by soils, Environ. Sci. Tech., 23: 1046-1056.

FÄRM, C. (2002): Metal sorption to natural filter substrate for storm water treatment-column studies. Sci. Total Environ., 298: 17-24.

FERGUSSON, J.E. (1990): The heavy elements: Chemistry, environmental impact an health effects: Pergamon Press, 614p.

FILELLA, M. & BUFFLE, J. (1993): Factors controlling the stability of submicron colloids in natural waters. Colloids and Surf. A., 73: 255-273.

FLORENCE, T.M.; LUMSDEN, B.J. & FARDY, J.J. (1983). Evaluation of some physicochemical techniques for the determination of the fraction of dissolved copper toxic to the marine diatom Nitzschia closterium. Annals. of Chimica. Acta, 151: 281.

FRAÚSTO DA SILVA, J.J.R. & WILLIAMS, R.J.P. (1991): The biological chemistry of the elements; the inorganic chemistry of life, Clarendon Press, Oxford, 561p.

FROST, R.R.; GRIFFIN, R.A. (1978): Exchange-Adsorption of Cu, Zn, and Cd from deionised water and leachate solutions by clay minerals. In: GRIFFIN, R.A. & SHIMP, N.F. (Eds.): Attenuation of pollutants in municipal landfill leachate by clay minerals. Illinois State Geological Survey, Environmental Geological Notes, Number 78: EPA-600/2-78-157.

FUCHS, S.; HARITOPOULOU, T.; SCHÄFER, M. & WILHELMI, M: (1997). Heavy metals in freshwater ecosystems introduced by urban rainwater runoff. Monitoring of suspended solids, river sediments and biofilms. Wat. Sci. Technol., 36: 277-282.

FULLER, C.C.; DAVIS, J.A. (1989): Influence of coupling of sorption and photosynthetic processes on trace element cycles in natural waters. Nature, 340: 152-154.

FULLER, W.H.; KORTE, N.E.; NIEBLA, E.E.; ALESII, B.A. (1976): Contribution of the soil to the migration of certain common and trace element. Soil Sci., 122: 223-235.

FURNESS, R.W. & RAINBOW, P.S. (1990): Heavy metals in the marine environment. CRC Press, Boca Raton, FL, 256p.

GAMBRELL, R.P.; WIESEPAPE, J.B.; PATRICK, W.H., JR. & DUFF, M.C. (1991): The effects of pH, redox, and salinity on metal release from contaminated sediment. Water, Air, and Soil Pollution, 57-58: 359-367.

GARRELS, R.M. & CHRIST, C.L. (1965): Solutions, minerals and equilibria, Harper & Row, New York, 216-222.

GEBEL, T. (2000): Confounding variables in the environmental toxicology of arsenic, Toxicology, 144: 155-162.

GIBBS, R.J. (1970): Mechanism controlling world water chemistry. Science, 170:1088-1090.

GIBSON, M.G. & FARMER, J.G. (1986): Multi-step chemical extraction of heavy metals from urban soils. Environ. Pollut. B., 11:117-135

GIRMAY, H.S. & ASSEFA, G. (1989): The Addis Ababa-Nazareth Volcanics: A Miocene-Pleistocene volcanic succession in Ethiopian Rift. SINET: Ethiop. J. Sci., 12: 1.

GOVINDARAJU, K. (1994): Compilation of working values and sample descriptions for 383 geostandards, Geostandards Newsletter, 18:1-158.

GOYER, R.A. (1993): Lead toxicity: Current concerns. Environ. Health Perspect., 100: 177-187.

GOYER, R.A. (1996): Toxic effects of metals. In: Casarett & Doull's Toxicology. The Basic Science of Poisons, 5th Edition, Klaassen, C.D. (Ed,): McGraw-Hill Health Professions Division, 848.

GRAN, G. (1952): Determination of the equivalence point in potentiometric titration. Part II. Analyst, 77: 661.

GRAEME, H.M. &. POLLACK, J.V.C. (1998): Selected topics: Toxicology: Part I. Arsenic and mercury. J. Emerg. Med. 16: 45-56.

GRIM, R.E. (1968): Clay Mineralogy. McGraw-Hill Book Company, New York. 384p.

GUSTAFSSON, O.; GSCHWEND, P.M. (1997): Aquatic colloids: Concepts, definitions, and current challenges. Limnol. Oceanogr., 42: 519.

GUSTAFSSON, J.P.; PECHOVA, P; BERGGREN, D. (2003): Modeling metal binding to soils: the role of natural organic matter. Environ. Sci. Technol., 37: 2767-2774.

GÜVEN, N. (1988); Smectites. In: BAILEY, S.W. (Ed.): Hydrous phyllosilicates (Exclusive Micas), Mineralogical Society of America, 497-559.

HAKANSON, L. & JANSSON, M. (1983): Principles of lake sedimentology. Springer Verlag, Berlin, 316p.

HALL, G.E.M.; GAUTHIER, G.; PELCHAT, J.C.; PELCHAT, P.; VAIVE, J.E. (1996): Application of a sequential extraction scheme to ten geological certified reference materials for the determination of 20 elements. J. Anal. Atom. Spectrom, 11: 787-796.

HARRISON, R.M.; LAXEN, D.P.H. & WILSON, S.J. (1981): Chemical association of lead, cadmium, copper, and zinc in street dust and roadside soil. Environ. Sci. Technol. 15: 1378-1383.

HART, B.T. (1982): Uptake of trace metals by sediments and suspended particulates: A review. Hydrobiology, 91: 299-313.

HAZEN, A. (1892): Physical properties of sands and gravels with reference to use in filtration, Report to Mass. State Board of Health, 539-556.

HE, X.T.; LOGAN, T.J.; TRAINA, S.J. (1995): Physical and chemical characteristics of selected U.S. municipal solid-waste composts. J. Environ. Qual., 24: 543-552.

HEIN, H.; SCHWEDT, G. (1995): Richt-und Grenzwerte Wasser-Boden-Abfall-Gefahrstoffe-Luft. 4. Überarbeitete und erhebliche erweiterte Auflage. Arbeitsmittel Umwelt Magazin. Vogel Verlag, Würzburg.

HELLAWELL, J.M. (1986): Biological indicators of freshwater pollution and environmental management. Elsevier Applied Science Publishers, London and New York, 546p.

HEM, J.D. (1985): Study and interpretation of the chemical characteristics of natural waters: (3rd Edition): Water Supply Paper, 2254, U.S. Geological Survey, 264pp.

HEM, J.D. (1989): Study and interpretation of the chemical characteristics of natural waters: (3rd Edition): U.S. Geological Survey Water Supply Paper, 2254, 263p.

HICKEY, M.G.; KITTRICK, J.A. (1984): Chemical partitioning of cadmium, copper, nickel and zinc in soils and sediments containing high levels of heavy metals. J. Environ. Qual., 13: 372-376.

HO, Y.S.; HUANG, C.T.; HUANG, H.W. (2002): Equilibrium sorption isotherm for metal ions on tree fern. Process Biochem., 37: 1421-1430.

HOLMSTROM, H.; LJUNGBERG, J.; OHLANDER, B. (1999): Role of carbonates in mitigation of metal release from mining waste. Evidence from Humidity Cells. Environ. Geol., 37(4): 267-280.

HONEYMAN, B.D. (1984): Cation and anion adsorption at the oxide/solution interface in systems containing binary mixtures of adsorbents: An investigation of the concept of adsorptive additivity, Ph.D. thesis, Stanford University, Palo Alto, CA (Diss. abstr. AAG8420552), 383p.

HONEYMAN, B.D. & SANTSCHI, P.H. (1988): Metals in aquatic systems: predicting their scavenging residence times from laboratory data remains a challenge, Environ. Sci. Technol., 22: 862.

HOROWITZ, A.J.; RINELLA, F.; LAMOTHE, P.; MILLER, T.; EDWARDS, T.; ROCHE, R. & RICKERT, D. (1990): Variations in suspended sediment and associated trace element concentrations in selected riverine cross sections. Environ. Sci. Technol., 24:1313.

HSU, P.H. (1989): Aluminum oxides and oxyhydroxides. In: DIXON, J.B. & WEED, S.B. (Eds.): Minerals in soil environments. ASA and SSSA, Madison, 331-378.

HUHEEY, J.E. (1972): Inorganic chemistry: Principles of structure and reactivity. 3rd Edition, Harper and Row, New York, 699p.

HUNTSMAN, S.A. & SUNDA W.G. (1980): The role of trace metals in regulating phytoplankton growth with emphasis on Fe, Mn and Cu. In: MORRIS, I. (Ed.): The physiological ecology of phytoplankton. Blackwell Scientific Publications, Oxford, England, 285-328.

INTERLANDI, S.J.; CROCKETT, C.S. (2003): Recent water quality trends in the Schuylkill River, Pennsylvania, USA: A preliminary assessment of the relative influences of climate, river discharge and suburban development. Water Res., 37: 1737-1748.

IRMER, U. (1997): Bedeutung Von Hintergrundwerten für Qualitätsanforderung an Oberflächenwassern. IKSE - Workshop: Bewertung der Ergebnisse aus der Elbeschadstofforschung. Geesthacht, 36-40.

ISMAIL, S.S. (1996): Distribution of trace elements in Egyptian ground and Nile water. J. Trace Microprobe Tech. 14 (1): 243-253.

ITANNA, F. (1998): Metal concentrations of some vegetables irrigated with industrial liquid waste at Akaki, Ethiopia. SINET: Ethiop. J. Sci., 21: 133-144.

ITANNA, F.; ANDERSON, D.; STAHR, K. (2003): Influence of soil type differences on the distribution of DTPA extractable heavy metals in soils irrigated with industrial effluents. SINET: Ethiop. J. Sci., 26 (1): 47-54(8).

JACOB, D.L; OTTE, M.L. (2004): Long-term effects of submergence and wetland vegetation on metals in a 90-year old abandoned Pb-Zn mine tailings pond. Environ. Pollut., 130: 337-345.

JANG, A.; SEO, Y.; BISHOP, P.L. (2005): The removal of heavy metals in urban runoff by sorption on mulch. Environ. Pollut., 133: 117-127.

JARDINE, P.M.; SPARKS, D.L. (1984): Potassium-Calcium exchange in a multireactive soil system: I. Kinetics. Soil Sci. Soc. Am. J., 48: 39-45.

JAUDON, P.; MASSIANI, C.; GALEA, J.; REY, J.; VACELET, E. (1989): Groundwater pollution by manganese-manganese speciation-application to the selection and discussion of an in situ groundwater treatment. Sci. of the Tot. Environ., 84: 69-183.

JOHNSTON, C.A. (1993): Mechanisms of wetland-water quality interaction. In: MOSHIRI, G.A. (Ed.): Constructed wetlands for water quality improvement. Lewis Publishers Boca Raton, Ann Arbor, London, Tokyo, 656p.

KELLY, J.; THORNTON, I. & SIMPSON, P.R. (1996): Urban geochemistry: A study of influence of anthropogenic activity on heavy metal content of soils in traditionally industrial and non-industrial areas of Britain. Appl. Geochem., 11: 363-370.

KENNISH, M.J. (1992): Ecology of estuaries: Anthropogenic effects, CRC Press, Boca Raton (FL), 494p.

KIM, N.D.; FERGUSSON, J.E. (1991): Effectiveness of a commonly used sequential extraction technique in determining the speciation of cadmium in soils. Sci. Total Environ., 105: 191-209.

KINNIBURGH, D.G.; JACKSON, M.L. & SYERS, J.K. (1976): Adsorption of alkaline earth, transition, and heavy metal cations by hydrous oxide gels of iron and aluminum. Soil Sci. Soc. Am. J., 40: 796-800.

KINNIBURGH, D.G. & JACKSON, M.L. (1981): Cation adsorption by hydrous metal oxide and clays. In: ANDERSON, M.A. & RUBIN, A.J. (Eds.): Adsorption of inorganics at solid-liquid interfaces. Ann Arbor Science, Ann Arbor, MI., 91-160.

KRAMAR, U. (1997): Advances in energy-dispersive X-ray fluorescence. J. Geochem. Exploit., 58:73-80.

KRAUSE, E. & ETTEL, V.A. (1989): Solubilities and stabilities of ferric arsenate compounds. Hydrometallurgy, 22: 311-337.

KRÜGER, A.; SCHNEIDER, B. (1994): Untersuchungen zu Bindungsformen von Schwermetallen in Böden des Bitterfelder Industriegebies Forschungsbericht 01/94; Universität Leipzig, FB Physik und Geowissenschaften.

LAGREGA, M.D.; BUCKINGHAM, P.L. & EVANS, J.C. (1994): Hazardous waste management, McGraw Hill, New York, 530p.

LACKOVIC, K.; WELLS, J.D.; JOHNSON, B.B.; ANGOVE, M.J. (2004): Modeling the adsorption of Cd(II) onto kaolinite and Muloorina illite in the presence of citric acid. J. Colloid Interface Sci., 270: 86-93.

LAKE, D.L.; KIRK, P.; LESTER, J. (1984): Fractionation, characterization, and speciation of heavy metals in sewage sludge and sludge amended soil. A review. J. Environ. Qual., 13: 175-183.

LAWA, (2000): Bund/Länder-Arbeitsgemeinschaft Wasser, Entwurf einer deutschen Modellverordnung zur Umsetzung der Richtlinie 76/464/EWG.

LAWS, E.A. (1981): Aquatic pollution. John Wiley and Sons, New York, 301-369.

LECKIE, J.O. (1978): Effect of adsorbed complexing ligands on trace metal uptake by metal oxides. Environ. Sci. Technol., 12: 1309-1315.

LECKIE, J.O.; BENJAMIN, M.M.; HAYES, K.; KAUFMAN, G. & ALTMAN, S. (1980): Adsorption/co-precipitation of trace elements from water with iron oxyhydroxides. EPRI-RP-910. Electric Power Research Institute, Palo Alto, CA, 245p.

LEE, G.; BIGHAM, J.M.; FAURE, G. (2002): Removal of trace metals by coprecipitation with Fe, Al and Mn from natural waters contaminated with acid mine drainage in the Ducktown Mining District, Tennessee. Appl. Geochem., 17: 569-581.

LEEUWANG, P. (1990): Ecotoxicology. Lecture Notes: Part 2. IHE, Delft, 2.

LELAND, J.V. & KUWABARA, J.S. (1985): In: RAND, G.M. & PETROCELLI, S.R. (Eds.): Fundamentals of aquatic toxicology. Taylor and Francis, New York, 374-415.

LIM-NUNEZ, R.; GILKES, R.J. (1987): Acid dissolution of synthetic metal-containing goethites and hematites. In: SCHULTZ, L.G.; VAN OLPHEN, H. & MUMPTON, F.A. (Eds.): Proceedings of the International Clay Conference, The Clay Minerals Society, Bloomington, Indiana, 197-204.

LINDSAY, W.L. (1979): Chemical equilibria in soils. John Wiley and Sons. New York, 449p.

LISTON, P. & MAHER, W. (1986): Trace metal export in urban runoff and its biological significance. Bull. Environ. Contam. Toxicol., 36: 900-905.

LUND, W. (1990): Speciation Analysis-Why and How? Fresenius' J. Anal. Chem., 337: 557-564.

LUOMA, S.N. (1989): Can we determine the biological availability of sediment-bound trace elements? Hydrobiologia, 176/177: 379-396.

LUOMA, S.N. (1983): Bioavailability of trace metals to aquatic organisms. A Review. Sci. Total Environ., 28: 1-22.

MADISON, R.J. & BRUNETT, J.O. (1984): Overview of the occurrence of nitrate in ground water of the United States. In: U.S. Geological Survey, National Water Summary. U.S. Geological Survey Water-Supply Paper 2275: 93-105.

MANAHAN, S.E. (1996): Fundamentals of environmental chemistry. In: BOULDING, J.R. (Ed.): EPA environmental assessment source book. Ann Arbor Press, Inc., 424p

MANCE, G. (1987). Pollution threat of heavy metals in aquatic environments. Pollution Monitoring Series, Elsevier Applied Science, London, 327p.

MARIN B.; VALLADON, M.; POLVE, M.; MONACO, A. (1997): Operational speciation of cadmium, copper, lead and zinc in the NIST Standard Reference Materials. Anal. Chim. Acta, 342: 91.

MARTIN, J.M.; MEYBECK, M. (1979): Elemental mass-balance of material carried by world major rivers. Mar. Chem., 7: 173-206.

MATAGI, S.V.; SWAI, D.; MUGABE, R. (1998): A review of heavy metal removal mechanisms in wetlands. Afr. J. Trop. Hydrobiol. Fish, 8: 23-35.

MATTIGOD, S.V.; RAI, D.; FELMY, A.R. & RAO, L. (1997): Solubility and solubility product of crystalline $Ni(OH)_2$. J. Sol. Chem., 26: 391-403.

MCBRIDE, M.B.; PINNAVAIA, T.J.; MORTLAND, M.M. (1975): Electron spin relaxation and the mobility of manganese (II) exchange ions in smectites. Am. Mineral, 60: 66-72.

MCBRIDE, M.B. (1989): Reactions controlling heavy metal solubility in soils. Adv. Soil Sci., 10:1-56

MCBRIDE, M.B. (1994): Environmental chemistry of soils. New York: Oxford Univ. Press, 585p.

MCGRATH, S.P. (1995): Chromium and nickel. In: ALLOWAY, B.J. (Ed): Heavy metal in soils. 2nd ed. Blackie Academic and Professional, 152-178.

MCLEAN J.E.; BLEDSOE, B.E. (1992): Behavior of metals in soils. Ground water issue. USEPA report, 25p.

MERTZ, W. & CORNAZER, T. (1971): Newer Trace elements in nutrition. Dekker, New York, 451p.

MEYBECK, M. (1982): Carbon, nitrogen and phosphorus transport by world rivers. Amer. J. Sci., 282: 282-450.

MEYBECK, M. (1988): How to establish and use world budgets of riverine materials. In: LERMAN, A. & MEYBECK, M. (Eds.): Physical and chemical weathering in geochemical cycles. Kluwer Academic Publishers, Dordrecht, 247-72.

MIRAGAYA, J.G.; SOSA, A.M. (1994): Trace metals in Valencia lake (Venezuela) sediments. Water Air Soil. Pollut., 77:141-150.

MISONO, M.; OCHIAI, E.; SAITO, Y.; YONEDA, Y. (1967): A new dual parameter scale for the strength of Lewis acids and bases with the evaluation of their softness. J. Inorg. Nucl. Chem., 29: 2685-2691.

MONSER, L.; ADHOUM, N. (2002): Modified activated carbon for the removal of copper, zinc, chromium, and cyanide from wastewater. Sep. Sci. Technol., 26(2-3): 137-146.

MOREL, F.M.M.; HERING, J.G. (1993): Principles and applications of aquatic chemistry. London: Wiley Interscience, 588p.

MOREL, F.M.M.; DZOMBAK, D.A.; PRICE, N.M. (1991): "Heterogeneous interactions in coastal waters." In: MANTOURA, R.F.C.; MARTIN, J.M. & WOLLAST, R. (Eds.): Ocean margin processes in global change. Wiley-Interscience, New York, 165-180.

MORTON, W.H.; REX, D.C.; MITCHELL, J.G.; MOHR, P.A. (1979): Rift ward younging of volcanic units in the Addis Ababa region, Ethiopian rift valley. Nature, 280: 284-288.

MULLER, G. (1988): Chemical decontamination of dredged materials, combustion residues, soil and other materials contaminated with heavy metals. In: WOLF, W.J. VAN DE BRINK & COLON, F.J. (Eds.): 2nd international TNO/BMFT conference on contaminated soil. Kluwer Dorrecht, 2: 1439-1442.

MULLER, B. & SIGG, L. (1990): Interaction of trace metals with natural particle surfaces: Comparison between adsorption experiments and field measurements. Aquat. Sci., 52: 75-92.

MULLER, F.; BESSON, G.; MANCEAU, A.; DRITS, V.A. (1997): Distribution of isomorphous cations within octahedral sheets in montmorillonite from Camp - Bertaux. Phys. Chem. Minerals, 24: 159-166.

MURRAY, K.S. (1996): Statistical comparisons of heavy-metal concentrations in river sediments. Environ. Geol., 27: 54-58.

NACHTEGAAL, M. & SPARKS, D.L. (2003): Nickel sequestration in a kaolinite-humic acid complex. Environ. Sci. Tech., 37: 529-534.

NAIDU, R.; SUMNER, M.E.; HARTER, R.D. (1998): Sorption of heavy metals in strongly weathered soils. An overview. Environ. Geochem. Health, 20: 5-9.

NAIDU, R.; BOLAN, N.S.; KOOKANA, R.S.; TILLER, K.G. (1994): Ionic-strength and pH effects on the sorption of cadmium and the surface charge of soils. Eur. J. Soil. Sci., 45: 419-429.

NEWMAN, M.C. & MCINTOSH, A.W. (1991): Metal ecotoxicology. Eds: Concepts and applications. Chelsea, Lewis Publishers, 424p.

NIES, D.H. (1999): Microbial heavy metal resistance, Appl. Microbiol. Biotechnol. 51: 730-750.

NOVOTNY, V. (2002): Water quality: diffuse pollution and watershed management. John Wiley and Sons, 303p.

NRC (1994): Alternatives for groundwater cleanup, National Research Council, National Academy Press, Washington, D.C, 315p.

NRIAGU, J.O.; LAWSON, G.; WONG, H.K.T. & CHEAM, V. (1996): Dissolved trace metals in Lakes Superior, Erie, and Ontario. Environ. Sci. Technol., 30: 178-187.

NRIAGU, J.O.; PACYNA, J.M. (1988): Quantitative assessment of worldwide contamination of air, water and soils by trace metals. Nature, 333: 134-139.

ODENSASS, M.; SCHROERS. S. (2000): Recommendations for performance and evaluation of column experiments in accordance with federal soil protection and refuse dump regulation. (Empfehlungen für die Durchführung und Auswertung von Säulenversuchen gemäß Bundes-Bodenschutz- und Altlastenverordnung (BBodSchV). Landesumweltamt Nordrhein-Westfalen. Merkblätter), Internet unter http://www.lua.nrw.de

OECD, (1981): Guidelines for testing of chemicals. Simultaneous Test - Aerobic sewage treatment: 303A. OECD, Paris, France, 28p.

OEHHA Office of Environmental Health Hazard Assessment. (2001): Public health goals for chemicals in drinking water: Nickel. California Environmental Protection Agency, http://www.oehha.ca.gov/water/phg/81701PHG.html.

O'REILLY, S.E. (2002): Lead sorption efficiencies of natural and synthetic Mn and Fe-oxides. Dissertation submitted to the faculty of the Virginia Polytechnic Institute, 53p

OSCARSON, D.W.; HUANG, P.M.; DEFOSSE, C.; HERBILLON, A. (1981): Oxidative power of Mn (IV) and Fe (III) oxides with respect to As (III) in terrestrial and aquatic environments. Nature, 291: 50-51.

OUKI, S.K.; NEUFELD, R.D.; PERRY, R. (1997): Use of activated carbon for the recovery of chromium from industrial wastewaters. J. Chem. Technol. Biotechnol., 70(1): 3-8.

PAPELIS, C.; HAYES, K.F. (1996): Distinguishing between interlayer and external sorption sites of clay minerals using x-ray absorption spectroscopy. Coll. Surf. A., 107: 89-96.

PARIDA, K.M.; SATAPATHY, P.K.; DAS, N. (1996): Studies on Indian Ocean Manganese Nodules IV adsorption of some bivalent heavy metal ions onto ferromanganese nodules. J. Colloid Interf. Sci., 181: 456-462.

PARKS, G.A. (1965): The Isoelectric points of solid oxides, solid hydroxides, and aqueous hydroxo complex systems. Chem. Rev., 65: 177-198.

PATRICK, W.H.; GAMBRELL, R.P. & KHALID, R.A. (1990): Physiochemical factors regulating solubility and bioavailability of toxic heavy metals in contaminated Dredged Sediments. Utrecht Plant Ecology News Report, 11:44-51.

PEARSON, R.G. (1963): Hard and soft acids and bases. J. Am. Chem. Soc., 85: 3533-3539.

PENDIAS, A.K.; PENDIAS, H. (1984): Trace Elements in Soils and Plants. Boca Raton: CRC Press, 245p.

PENDIAS A.K.; PENDIAS, H. (2001): Trace elements in soils and plants. 3rd ed. CRC Press, 413p.

PETERS, W.R.; KHU, Y. (1988): The effect of tartrate, a weak complexing agent, on the removal of heavy metals by sulfide and hydroxide precipitation. Part. Sci. Technol., 6: 421-439.

PIERCE, M.L. & MOORE, C.B. (1982): Adsorption of arsenite and arsenate on amorphous iron hydroxide. Water Res., 16: 1247-1253.

PIRKLE, J.L.; KAUFMAN, R.B.; BRODY, D.J.; HICKMAN, T.; GUNTER, E.W. & PASCHAL, D.C. (1998): Exposure of the U.S. population to lead, 1991-1994. Environ. Health Perspect., 11: 745-750

PONNAMPERUMA, F.N.; LOY, T.A.; TIANCO, E.M. (1969): Redox equilibria in flooded soils: II. The manganese oxide systems. Soil Sci., 108: 48-57.

POST, J.E. (1992): Crystal structures of manganese oxide minerals. In: SKINNER, H.C.W. & FITZPATRICK, R.W. (Eds.): Biomineralization processes of iron and manganese: Modern and ancient environments. Catena Verlag, 21: 51-73.

POST, J.E. (1999): Manganese oxide minerals: Crystal structures and economic and environmental significance. Proc. Natl. Acad. Sci. U.S.A, 96(7): 3447-3454.

QUANTIN, C.; BECQUER, T.; ROUILLER, J.H.; BERTHELIN J. (2001): Oxide weathering and trace metal release by bacterial reduction in a New Caledonia Ferralsol. Biogeochemistry, 53: 323-340.

RAI, D; ZACHARA, J.M.; SCHWAB, A.P.; SCHMIDT, R.L.; GIRVIN, D.C. & ROGERS, J.E. (1984): Chemical attenuation rates, coefficients, and constants in leachate migration. Volume 1: A Critical Review. EA-3356, Electric Power Research Institute, Palo Alto, CA.

RAI, D.; ZACHARA, J.M.; SCHWAB, A.P. (1986): Manganese. In: Chemical attenuation rates, coefficients, and constants in leachate migration. Volume 1: A critical review. Report to Electric Power Research Institute, Palo Alto, CA, by Battelle, Pacific Northwest Laboratories, Richland, WA, 15-1-15-4.

RAMOS, L.; HERNANDEZ, L.M.; GONZALES, M.J. (1994): Sequential fractionation of copper, lead, cadmium and zinc in soils from Donana National Park. J. Environ. Qual., 23: 50-57.

RAMOS, R.L.; RANGEL-MENDEZ, J.R.; MENDOZA-BARRON, J.; FUENTES-RUBIO, L.; GUERRERO-CORONADO, R.M. (1997): Adsorption of cadmium (II) from aqueous solution onto activated carbon, Water Sci. Technol., 35: 205-211(7)

RAMSEY, M.H. (1993): Sampling and analytical quality control (SAX) for improved error estimation in the measurement of Pb in the environment using robust analysis of variance. App. Geochem., 2: 149-153.

RAURET, G.; LÓPEZ SÁNCHEZ, J.F.; SAHUQUILLO, A.; RUBIO, R.; DAVIDSON, C.M.; URE, A.M.; QUEVAUVILLER, PH. (1999). Improvement of the BCR three-step sequential extraction procedure prior to the certification of new sediment and soil reference materials. Environ. Monit., 1: 57-61.

RAURET, G.; LOPEZ-SACHEZ, J.F.; SAHUQUILLO, A.; BARAHONA, E.; LACHICA, M.; URE, A.M.; DAVIDSON, C.M.; GOMEZ, A.; LUCK, D.; BACON, J.; YLI-HALLA, M.; MUNTAU, H. & QUEVAUVILLER, PH. (2000): Application of a modified BCR sequential extraction (three-step) procedure for the determination of extractable trace metal contents in a sewage sludge amended soil reference material (CRM 483), complemented by a three-year stability study of acetic acid. J. Environ. Monit., 2: 228-233.

RAVEN, K.P.; JAIN, A.; LOEPPERT, R.H. (1998): Arsenite and arsenate adsorption on ferrihydrite: Kinetics, equilibrium, and adsorption envelopes. Environ. Sci. and Technol., 32: 344-349.

RICORDEL, S.; TAHA, S.; CISSE, I.; DORANGE, G. (2001). Heavy metals removal by adsorption onto peanut husks carbon: characterization, kinetics study and modeling. Sep. Purif. Technol., 24: 389-401.

SALBU, B. (1995): Trace elements in natural waters. CRC Press, 160p.

SARMANI, S.; ABDULLAH, M.P.; BABA, I.; MAJID, A.A. (1992): Inventory of heavy metals and organic micro pollutants in an urban water catchment drainage basin. Hydrobiol., 235/236: 669-674.

SCHECKEL, K.G. & SPARKS, D.L. (2000): Kinetics of the formation and dissolution of Ni precipitates in a gibbsite/amorphous silica mixture. J. Colloid Interface Sci., 229: 222-229.

SCHECKEL, K.G. & SPARKS, D.L. (2001): Temperature effects on nickel sorption kinetics at the mineral-water interface. Soil Sci. Soc. America J., 65: 719-728.

SCHLEGEL, M.L.; MANCEAU, A.; CHARLET, L. (1999b): Sorption of metal ions on clay minerals: 2. Mechanism of Co sorption on hectorite at high and low ionic strength and impact on the sorbent stability. J. Colloid Interface Sci., 220: 392-405.

SCHWERTMANN, U.; TAYLOR, R.M. (1989): Iron oxides. In: DIXON, J.B. & WEED, S.B. (Eds.): Minerals in soil environments. ASA and SSSA, Madison, WI., 379-438.

SCHWERTMANN, U.; CORNELL, R.M. (1991): Iron oxides in the laboratory. Preparation and characterization. 2nd Edition, Wiley VCH, Weinheim, 236p.

SCOTT, D.T.; MCKNIGHT, D.M.; VOELKER, B.M.; HRNCIR, D.C. (2002): Redox processes controlling manganese fate and transport in a mountain stream. Environ. Sci. Techno., 36(3): 453-459.

SCOTT, M.J.; MORGAN, J.J. (1995): Reactions at oxide surfaces. 1. Oxidation of As(III) by synthetic birnessite. Environ. Sci. Technol., 29: 1898-1905.

SEURECA, (1991): Feasibility study and preliminary design, detailed report; Volume 4 Groundwater resource, Addis Ababa, AAWSA.

SHIM, J.W.; PARK, S.J.; RYU, S.K. (2001): Effect of modification with HNO_3 and NaOH by pitch-based activated carbon fibers. Carbon, 39: 1635-1642.

SINCLAIR, P.; BECKETT, R. & HART, B.T. (1989): Trace elements in suspended particulate matter from the Yarra River, Australia. Hydrobiologia, 177: 239-248.

SINGH, M.; ANSARI, A.A.; MÜLLER G. & SINGH, I.B. (1997): Heavy metals in freshly deposited sediments of the Gomati River (a tributary of the Gangs River): effects of human activities. Environ. Geol., 29: 247-252.

SINGH, B.; GILKES, R.J. (1992): Properties and distribution of iron oxides and their association with minor elements in the soils of southwestern Australia. J. Soil Sci., 43: 77-98.

SMIEJAN, A.; WILKINSON, K.J; ROSSIER, C. (2003): Cd bioaccumulation by a freshwater bacterium, Rhodospirillum rubrum. Environ. Sci. Technol., 37: 701-706.

SMITH, L.A.; MEANS, J.L.; CHEN, A.; ALLEMAN, B.; CHAPMAN, C.C.; TIXIER, J.S., JR.; BRAUNING, S.E.; GAVASKAR, A.R. & ROYER, M.D. (1995), Remedial options for metals-contaminated sites, Lewis Publishers, Boca Raton, FL, 221p.

SOON, Y.K. & BATES, T.E. (1982): Chemical pools of cadmium, nickel, and zinc in polluted soils and some preliminary indications of their availability to plants. J. Soil Sci., 33:477- 488.

SPARKS, D.L. (1995): Environmental soil chemistry. San Diego, CA: Academic Press, 267p.

SPOSITO, G.P. (1983): The chemical forms of trace metals in soils. In: THORNTON, I. (Ed): Applied Environmental Geochemistry. Academic Press, London, 123-170.

SPOSITO, G.P. (1984): The surface chemistry of soils. New York: Oxford Univ. Press, 234p.

SPOSITO, G.P. (1989): The chemistry of soils. Oxford University Press, US. Technology & Industrial Arts, 304p.

STOKES, P.M.; CAMPBELL, P.G.C.; SCHROEDER, W.H.; TRICK, C.; FRANCE, R.L.; PUCKETT, K.J.; LAZERTE, B.; SPEYER, M.; HANNA, J.E.; DONALDSON, J. (1988): Manganese in the Canadian environment. Ottawa, Ontario, National Research Council of Canada, Associate Committee on Scientific Criteria for Environmental Quality (NRCC No. 26193), 30-32.

STONE, M.; DROPPO, I.G. (1996). Distribution of lead, copper and zinc in size-fractionated riverbed sediment in two agricultural catchments of southern Ontario, Canada. Environ. Pollut., 93: 353.

STRIEGL, R.G. (1987): Suspended sediment and metals removal from urban runoff by a small lake. Wat. Resources Bull., 23: 985-996.

STUMM, W.; MORGAN, J.J. (1981): Aquatic chemistry: an introduction emphasizing chemical equilibria in natural waters, Wiley-Interscience, Toronto, 357p.

STUMM, W.; MORGAN, J.J. (1996): Aquatic chemistry. 3rd ed. John Wiley & Sons, New York, 1022p.

SUN, X.; DONER, H.E. (1996): An investigation of arsenate and arsenite bonding structures on goethite by FTIR. Soil Sci., 161: 865-872.

SUN, X.; DONER, H.E. (1998): Adsorption and oxidation of arsenite on goethite. Soil Sci., 163: 278-287.

SUTHERLAND, R.A.; TACK, F.M.G; TOLOSA, C.A.; VERLOO, M.G. (2000): Operationally defined metal fractions in road deposited sediment, Honolulu, Hawaii. J. Environ. Qual., 29: 1431-1439.

SVERJENSKY, D.A. (1994): Zero-point-of-charge prediction from crystal-chemistry and solvation theory. Geochim. Cosmochim. Acta, 58: 3123-3129

TADESSE, Y. (2004): Groundwater Modeling. A Case Study on Volcanic Water Supply Aquifer (Akaki Well field) of the city of Addis Ababa, 21p.

TANG L.; SPARKS, D.L. (1993): Cation-exchange kinetics on montmorillonite using pressure-jump relaxation. Soil Sci. Soc. Am. J., 57: 42-46.

TAYLOR, R.M. (1986): Non-silicate oxides and hydroxides. In: NEWMAN, A.C. (Ed.): Clay Minerals. Telford, London, 129-185.

TAYLOR, S.R.; MCLENNAN, S.M. (1985): The continental crust: its composition and evolution. Blackwell, Oxford, 312p.

TERTIAN, R.; CLAISSE, F. (1982): Principles of Quantative X-ray Fluorescence Analysis. Heyden & Son Ltd, 385p.

TESSIER, A.; CAMPBELL, P.G.C.; BISSON, M. (1979). Sequential extraction procedure for the speciation of particulate trace metals. Anal. Chem., 51: 844.

THORNTON, I. (1991): Metal contamination of soils in urban areas. In: BULLOCK P., GREGORY P.J. (Eds.): Soils in the Urban Environment. Blackwell, 47-75.

TOLERA, H.H. (2005): Characterization of contaminant levels in surface and groundwater of the Akaki Sub-Basin (unpublished report). Institute of Mineralogy and Geochemistry, University of Karlsruhe, 11p.

TSEHAYU, K. & HAILEMARIAM, T. (1990): Engineering Geological Mapping of Addis Ababa. Ethiopian Institute of Geological Survey.

TSIPURSKY, S.I. & DRITS, V.A. (1984): The distribution of octahedral cations in the 2:1 layers of dioctahedral smectites studied by oblique texture electron diffraction. Clay Miner., 19: 177-193.

ULRICH, B. (1991): An ecosystem approach to soil adification. In: ULRICH, B.; SUMMER, ME. (Eds.): Soil Adicity. Berlin Heidelberg New York, 28-79.

URE, A.M.; DAVIDSON, C.M. (2001): Chemical speciation in the environment, Blackie, Glasgow, 265-321.

USEPA, (1999): National Secondary Drinking Water Regulations. Secondary Maximum Contaminant Levels, 54p.

USPHS, (1997): Toxicological profile for lead on CD-ROM. Agency for Toxic Substances and Disease Registry. U.S. Public Health Service, 35p.

VAN DEN BERG, C.M.G. (1982): Determination of copper complexation with natural organic ligands in seawater by equilibration with MnO_2. II, Experimental procedures and application to surface seawater. Marine chem., 11: 323-342.

VEERESH, H.; TRIPATHY, S; CHAUDHURI, D; HART, B.R.; POWELL, M.A. (2003): Competitive adsorption behaviour of selected heavy metals in three soil types of India amended with fly ash and sewage sludge. Environ. Geol., 44: 363-370.

VERBURG, K.; BAVEYE, P.; MCBRIDE, M.B. (1995): Cation exchange hysteresis and dynamics of formation and breakdown of montmorillonite quasi-crystals. Soil Sci. Soc. Am. J., 59: 1268 -1273.

VERNIER, A. CHERNET, T. (1985): Hydrogeology of the Addis Ababa Area.

WANG T.S.; KUO, C.F.; JAN, K.Y. & HUANG, H. (1996): Arsenite induces apoptosis in Chinese hamster ovary cells by generation of reactive oxygen species, J. Cell. Physiol., 169: 256-268.

WARREN, B.E. (1981): X-Ray Diffraction. Courier Dover Publications, Science, 381p.

WEBSTER, J.R.; GOLLADAY, S.W.; BENFIELD, E.F.; D'ANGELO, D.J.; PETERS, G.T. (1990): Effects of forest disturbance on particulate organic matter budgets of small streams. J. N. Am. Benthol. Soc., 9(2): 120-140.

WEAST, R.C. (1986): CRC Handbook of Chemistry and Physics, 66th edition, CRC press, Inc. Boca Raton, Florida, B104p.

WEIGERT, P. (1991): Metal loads of food of vegetable origin including mushrooms. In: MERIAN, E. (Ed.): Metals and their compounds in the environment: Occurrence, analysis and biological relevance. Weinheim: VCH, 458-68.

WEINER, E.R. (2000): Application of environmental chemistry: a practical guide for environmental professionals. Lewis Publishers, CRC Press LLC, 40-41.

WHITFIELD, M. & TURNER D.R. (1987): The role of particles in regulating the composition of seawater. In: STUMM, W., (Ed.): Aquatic surface chemistry, Chemical processes at the particle-water interface. Wiley-Interscience Series on Environmental Science and Technology, 457-494.

WHITTON, B.A. & SAY, P.J. (1975): Heavy metals. In: WHITTON, B.A. (Ed.): River ecology. Blackwell Scientific Publications, Oxford, 286-312.

WILLIAMS, K.L. (1987): An Introduction to X-Ray Spectrometry: X-Ray Fluorescence and Electron Microprobe Analysis. Allen & Unwin, London, 370p.

WORKU, Y.; EJIGU, S.; ERGE, W.; JEMANEH, L. (1999): Chemical, physical, and microbiological characteristics of various sources of water in and around Addis Ababa. Ethiop. J. Health Dev., 13: 239-246.

WYMAN, S.A. (1993): The potential for heavy metal migration from sediments of Lake Coeur D'alene into the Rathdrum Prairie aquifer Kootenai County, Idaho. Research Technical Completion Report. Idaho Water Resources Research Institute, University of Idaho, 158p.

XIAN, X. (1989): Effect of chemical forms of cadmium, zinc, and lead in polluted soils on their uptake by cabbage plants. Plant Soil, 113: 257-264.

YAMADA, I. & SHOJI, S. (1975): Relationship between particle size and mineral composition of volcanic ashes. Tohoku. J. Agr. Res., 26: 7-10.

TADESSE, Y. (2004): Groundwater modeling. A case study on volcanic water supply aquifer /Akaki well field/ of the city of Addis Ababa, 21p.

ZAW, M. & CHISWELL, B. (1999): Iron and manganese dynamics in lake water. Water Res., 33: 900-1910.

ZEIEN, H. & BRÜMMER, G.W. (1989): Chemische Extraktion zur Bestimmung von Schwermetallbindungsformen in Böden. (In German.) Mitteilgn. Dtsch. Bodenkundl. Gesellsch., 59: 505-510.

Appendix 1. Water parameters, concentrations of cation and anions of surface and groundwater samples in the Akaki Sub-basin. April 2003.

Sample	O_2 [mgl^{-1}]	T °C	pH	Conductivity [μScm^{-1}]	Na^+ [mgl^{-1}]	K^+ [mgl^{-1}]	Mg^{2+} [mgl^{-1}]	Ca^{2+} [mgl^{-1}]	Cl^- [mgl^{-1}]	NO_3^- [mgl^{-1}]	SO_4^{2-} [mgl^{-1}]
Ku1	0.26	21	6.90	437	62.00	26.00	15.00	37.75	31.35	1.84	17.29
De2	2.13	29	7.28	676	86.00	18.00	13.00	39.50	36.87	3.78	91.13
De3	3.63	25	6.70	410	6.00	3.00	14.00	43.50			
Dk4	2.86	22	7.19	756	50.00	29.00	16.00	44.00	54.62	0.00	40.10
Dk5	2.42	21	7.44	611	46.00	15.00	14.00	40.00	45.57	1.02	28.16
De6	3.54	30	7.33	678	82.00	9.00	13.00	47.75	39.48	2.79	91.11
Gr7	8.49	29	8.00	523	14.00	47.00	23.00	27.50	10.88	0.00	7.40
Be8	5.64	22	7.30	773	19.00	3.00	22.50	38.50	5.43	13.46	6.20
Sd9	5.55	21	7.58	625	26.00	4.00	25.00	49.50	15.76	43.26	17.31
E10	4.97	30	6.90	585	27.00	3.00	24.00	49.00	7.02	21.18	10.96
Ju11	3.25	19	7.32	516	38.00	20.00	40.00	26.00	46.58	14.81	43.66
Tl12	2.17	22	6.98	563	66.00	10.00	39.00	55.00	4.80	5.67	9.42
Iv13	0.77	25	6.60	852	28.00	6.00	42.00	74.00	34.33	18.68	36.98
Akt14	0.4	26	7.29	1029	68.00	5.00	19.50	70.50	57.64	0.00	71.87
Ah15	1.37	22	7.01	866	28.00	5.00	21.00	73.00	36.37	19.17	36.02
Ls16	3.82	25	7.35	609	84.00	7.00	22.50	42.00	5.04	8.58	10.14
Ak17	5.4	20	7.56	560	50.00	6.00	23.00	30.00	8.41	11.34	12.20
El18	0.04	21	7.45	1058	86.00	40.00	20.00	56.00			
El19	0.06	27	6.79	1037	29.00	4.00	24.50	32.00			
Td20	6.34	23	7.16	528	20.00	3.00	32.00	32.50	6.82	20.79	10.30
Am21	5.65	23	7.15	475	25.00	4.00	16.00	43.50	5.45	13.23	63.30
Ab22	5.26	23	7.29	512	11.00	3.00	8.50	21.25	5.37	16.38	8.05

Water parameters of surface and groundwater samples in the Akaki Sub-basin.

October 2005

Sample		O2 (mg/l)	T (℃)	pH	Conductivity (µS/cm)
Sd9	Sidamo Awash	5.81	22.30	7.1	489
E10	EEPCO	6.00	22.00	7.2	511
Ju11	Small Akaki (Jugan)	5.94	19.50	7.7	465
T12	Tilinto	5.63	21.60	7.3	580
Iv13	Indian Village	6.20	20.00	7.1	727
Ah15	Aba Hana	6.36	19.20	7.1	830
Ls16	Lesperance	5.84	28.00	7.5	727
Ak17	Akaki Metal Factory	6.58	20.30	7.8	554
El18W	ELICO Tannery	1.55	20.70	7.2	762
Akk	Kality Metals	4.34	22.20	7.6	548
Gx	Garment Express	6.50	25.90	7.6	469

Appendix 2. Dissolved heavy metal concentrations (µg/l) of surface and groundwater in the Akaki Sub-basin. (April 2003)

Sample	45Sc	51V	52Cr	55Mn	56Fe	59Co	60Ni	65Cu	66Zn	75As	85Rb
Ku1	0.05	6.22	0.36	352	132	1.66	5.46	2.89	174	1.40	11.4
De2	0.02	0.61	0.08	23	13	0.02	0.43	0.58	117	0.39	44.0
De3	0.03	8.55	0.44	2	8	0.03	0.75	2.35	616	1.42	29.1
Dk4	0.02	2.93	0.41	2835	98	3.80	9.70	2.14	43	2.75	16.0
Dk5	0.01	6.33	0.37	1153	27	2.33	7.32	2.70	24	1.73	14.1
De6	0.02	0.70	0.06	35	16	0.03	0.41	0.89	77	0.40	50.8
Gr7	0.06	5.54	0.58	4833	379	21.16	45.54	13.72	1336	1.33	18.6
Be8	0.04	40.62	4.17	14	4	0.11	0.99	3.97	327	0.61	9.4
Sd9	0.04	29.97	1.50	18	3	0.16	0.68	3.01	71	0.88	11.1
E10	0.04	45.41	1.66	2	3	0.06	1.06	3.45	1623	0.34	4.8
Ju11	0.03	5.95	0.57	834	87	4.38	8.30	5.93	54	1.29	13.1
Tl12	0.05	10.07	0.10	10	40	0.10	0.78	3.46	2734	0.81	18.7
Iv13	0.03	12.92	1.21	9	7	0.13	1.72	4.95	886	0.55	15.9
Akt14	0.04	8.94	0.89	159	63	0.26	3.13	4.73	28	0.73	22.8
Ah15	0.03	14.75	2.03	1	8	0.11	1.61	1.24	14	0.61	14.2
Ls16	0.03	13.82	0.36	1	2	0.03	0.37	2.13	3194	1.05	18.3
Ak17	0.02	23.56	0.63	1	8	0.02	0.25	0.75	225	0.74	12.4
El18	0.03	4.50	1.64	1268	868	2.49	5.29	5.08	29	2.13	26.2
Td20	0.02	34.38	1.95	15	5	0.05	0.40	1.17	104	0.67	9.7
Am21	0.02	31.66	3.28	2	2	0.03	0.47	0.93	53	0.50	9.0
Ab22	0.02	36.82	2.39	1	1	0.02	0.49	1.03	42	0.60	10.1

Dissolved heavy metal concentrations (µg/l) of surface and groundwater in the Akaki Sub-basin (April 2003). *Contd.*

Sample	88Sr	98Mo	107Ag	111Cd	118Sn	121Sb	137Ba	205Tl	207Pb	209Bi	232Th	238U
Ku1	220.30	2.28	0.069	0.034	0	0.74	95.70	0.03	0.191	<0,04	<0,04	0.77
De2	251.60	18.87	0.014	0.038	<0,08	0.19	42.57	0.02	0.079	<0,04	<0,04	0.34
De3	283.90	2.11	0.009	0.018	<0,08	0.22	26.31	0.10	0.853	<0,04	<0,04	2.90
Dk4	268.30	3.98	0.015	0.020	<0,08	0.41	121.20	0.04	0.135	<0,04	<0,04	0.93
Dk5	271.00	2.96	0.008	0.017	<0,08	0.39	120.30	0.03	0.090	<0,04	<0,04	1.23
De6	258.40	18.60	0.006	0.031	<0,08	0.12	41.09	0.03	0.173	<0,04	<0,04	0.36
Gr7	266.40	6.26	0.039	0.120	0	0.30	346.60	0.04	0.452	<0,04	<0,04	0.81
Be8	394.80	1.85	0.004	0.021	<0,08	0.10	16.30	0.03	0.800	<0,04	<0,04	1.74
Sd9	538.40	1.32	0.004	0.010	<0,08	0.09	34.56	0.04	0.236	<0,04	<0,04	2.38
E10	517.20	1.52	0.006	0.059	<0,08	0.09	83.91	0.08	2.453	<0,04	<0,04	3.50
Ju11	160.30	3.56	0.011	0.044	0	1.48	96.87	0.03	0.554	<0,04	<0,04	0.72
Tl12	434.90	1.75	0.007	0.053	<0,08	0.25	13.77	0.05	4.712	<0,04	<0,04	4.76
Iv13	732.10	1.44	0.005	0.017	<0,08	0.08	19.68	0.05	3.391	<0,04	<0,04	7.84
Akt14	722.10	1.01	0.011	0.043	<0,08	0.16	25.50	0.05	0.218	<0,04	<0,04	5.46
Ah15	748.70	1.35	0.006	0.015	<0,08	0.10	12.10	0.05	0.182	<0,04	<0,04	6.62
Ls16	471.00	1.93	0.006	0.528	<0,08	0.12	9.79	0.13	7.435	<0,04	<0,04	4.43
Ak17	395.00	3.01	0.004	0.035	<0,08	0.06	21.76	0.08	1.806	<0,04	<0,04	5.12
El18	250.00	1.29	0.020	0.074	<0,08	0.34	140.10	0.11	0.987	<0,04	<0,04	0.66
Td20	318.00	2.16	0.002	0.023	<0,08	0.07	22.66	0.12	0.521	<0,04	<0,04	4.42
Am21	269.00	1.66	0.005	0.024	<0,08	0.07	14.52	0.10	0.735	<0,04	<0,04	2.92
Ab22	295.00	1.76	0.004	0.039	<0,08	0.04	14.63	0.08	0.381	<0,04	<0,04	3.09

Dissolved heavy metal concentrations (µg/l) of surface and groundwater in the Akaki Sub-basin. (October 2005)

** Bulk water samples not filtered through 0.45 µm pore size filter. El18W represents groundwater well at ELICO.*

Sample	45Sc	51V	52Cr	55Mn	56Fe	59Co	60Ni	65Cu	66Zn	75As	85Rb
Sd9	0.01	47.77	2.66	0.75	1.38	0.02	0.61	6.81	106.80	0.47	7.61
Sd9*	0.01	44.81	2.55	0.28	1.69	0.02	0.49	7.17	95.47	0.43	7.58
E10	0.01	47.13	2.31	0.64	1.65	0.02	0.69	2.04	591.70	0.60	8.70
E10*	0.01	44.97	2.29	0.31	1.25	0.02	0.36	1.58	607.50	0.56	9.23
Ju11	0.01	2.29	15.91	658.80	197.40	1.98	3.58	2.31	13.21	1.22	7.83
Ju11*	0.01	1.45	6.44	20.52	24.64	1.49	2.77	1.46	4.88	0.85	7.80
T12	0.01	9.74	0.11	1.75	20.00	0.02	0.80	2.95	276.00	0.80	14.36
T12*	0.01	8.69	0.10	1.93	22.10	0.02	0.48	1.94	265.10	0.67	13.25
Iv13	0.00	8.28	0.15	9.20	2.37	0.07	1.36	1.60	93.87	0.72	13.68
Iv13*	0.00	8.40	0.13	2.59	3.84	0.03	0.32	1.33	82.34	0.60	13.45
Ak14	0.01	8.80	1.05	113.80	268.30	0.21	1.74	2.76	44.07	0.28	34.74
Ak14*	0.00	5.09	0.65	66.34	118.10	0.15	1.23	2.40	28.12	0.30	19.24
Ah15	0.00	15.77	1.27	2.03	5.75	0.11	2.83	6.06	22.21	1.32	11.72
Ah15*	0.01	15.64	1.19	0.73	8.65	0.10	1.73	12.01	12.80	0.60	12.26
Ls16	0.01	50.13	2.46	2.69	2.75	0.03	0.79	2.68	511.40	0.55	8.46
Ls16*	0.01	48.69	2.47	0.41	2.81	0.02	0.40	2.93	489.10	0.44	8.79
Ak17	0.01	14.16	0.61	1.47	2.50	0.01	0.50	1.19	45.82	1.22	16.84
Ak17*	0.00	14.44	0.69	1.04	5.89	0.01	0.43	1.56	45.88	1.29	17.39
El18W	0.00	11.03	0.82	820.70	8.71	0.37	3.01	3.19	52.72	2.60	5.04
El18W*	0.00	11.50	1.87	836.20	39.55	0.37	2.50	3.05	49.11	2.35	4.98
Akk	0.02	37.34	1.28	0.86	6.34	0.02	0.61	2.12	253.30	0.59	12.54
Akk*	0.01	26.20	0.18	0.19	1.04	0.02	0.48	1.44	74.62	0.47	11.89
Gx	0.00	59.49	2.87	2.00	3.72	0.07	1.52	2.57	1186.00	0.40	7.37
Gx*	0.00	61.05	2.95	1.24	7.32	0.07	0.89	1.54	1206.00	0.42	7.66

Dissolved heavy metal concentrations (µg/l) of surface and groundwater in the Akaki Sub-basin (October 2005). *Contd.*

Sample	88Sr	98Mo	111Cd	121Sb	137Ba	205Tl	207Pb	232Th	238U
Sd9	389.2	1.64	0.05	0.13	23.45	0.01	0.28	0.01	1.38
Sd9*	398.4	1.33	0.02	0.14	24.92	0.01	0.21	0.01	1.36
E10	404.3	1.60	0.05	0.30	13.53	0.01	0.35	0.01	1.82
E10*	447.9	1.81	0.02	0.24	14.62	0.01	0.24	0.01	1.81
Ju11	166.1	0.85	0.05	0.32	68.11	0.01	0.48	0.02	0.33
Ju11*	98.57	1.19	0.03	0.29	32.43	0.01	0.11	0.05	0.20
T12	492	1.71	0.06	0.24	13.08	0.01	0.46	0.01	3.99
T12*	515.2	1.62	0.02	0.23	14.47	0.01	0.47	0.01	4.01
Iv13	631.3	1.24	0.25	0.23	17.81	0.01	0.37	0.03	3.90
Iv13*	622.9	1.46	0.01	0.18	16.97	0.01	0.43	0.03	4.95
Ak14	467.1	0.43	0.02	0.12	27.34	0.00	0.37	0.08	1.54
Ak14*	276	0.34	0.01	0.11	16.69	0.00	0.20	0.05	0.92
Ah15	823.8	2.32	0.18	0.16	15.59	0.01	0.24	0.01	6.64
Ah15*	812	1.75	0.02	0.14	13.06	0.01	0.34	0.01	8.60
Ls16	417	1.82	0.27	0.18	11.95	0.00	2.92	0.01	2.08
Ls16*	399.8	2.00	0.17	0.19	11.53	0.00	3.32	0.01	2.16
Ak17	413.9	4.00	0.02	0.21	12.80	0.01	0.19	0.01	3.71
Ak17*	443.3	4.09	0.02	0.17	13.90	0.01	0.17	0.01	3.85
El18W	300.6	1.60	0.05	0.50	155.30	0.02	0.27	0.02	0.90
El18W*	301.2	1.50	0.03	0.41	119.10	0.02	0.35	0.01	0.86
Akk	522.3	2.10	0.05	0.13	24.55	0.01	0.82	0.01	3.15
Akk*	558.1	2.65	0.01	0.13	22.85	0.01	0.09	0.01	0.30
Gx	321.7	1.82	0.05	0.18	17.16	0.01	1.09	0.00	1.49
Gx*	312.4	1.53	0.03	0.16	16.42	0.01	1.21	0.00	1.44

Comparison of concentrations of selected heavy metals between water samples collected (a) in April 2003 and (b) October 2005. Samples compared are Sd9 – El18. Samples from Ku1 – Be were not available during the October 2005 sampling campaign.

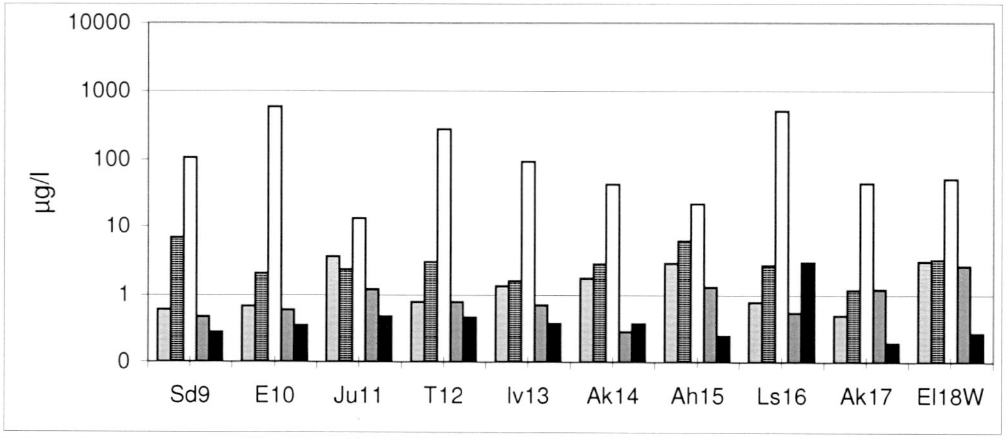

a) April 2003

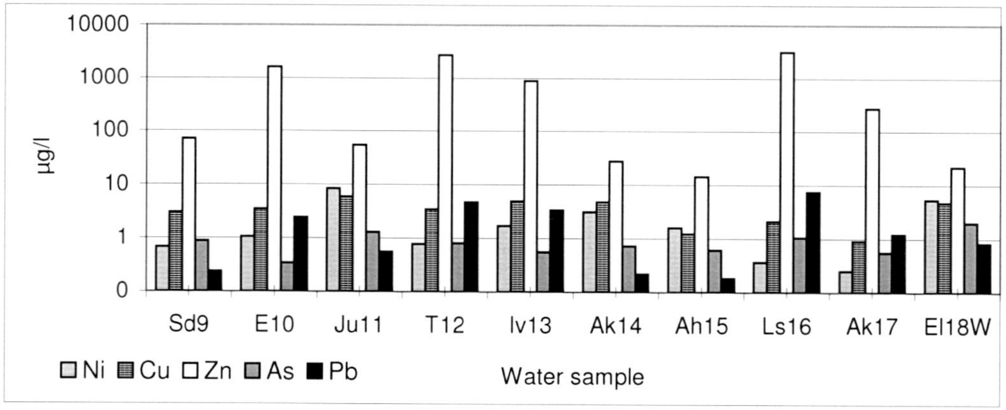

b) October 2005

Comparison of dissolved metal concentrations in samples collected during 2003 vs 2005 in relation to US EPA MCL standards for heavy metals and arsenic.

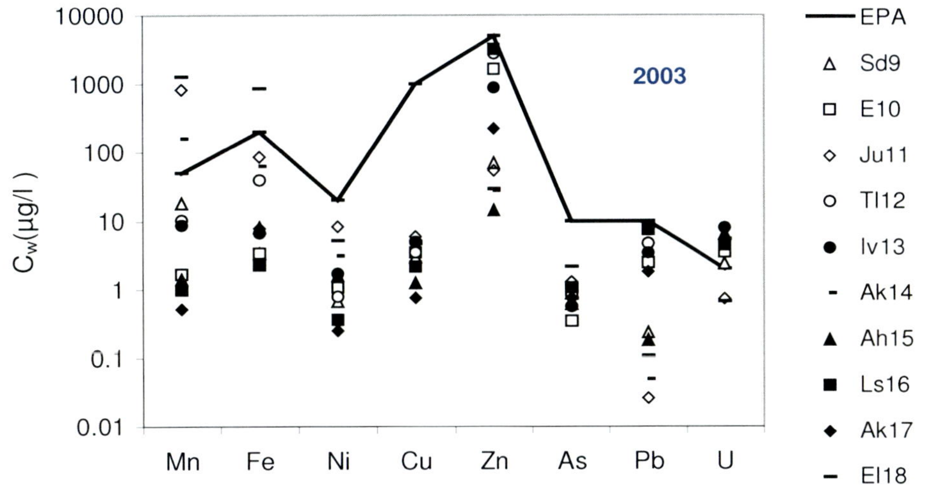

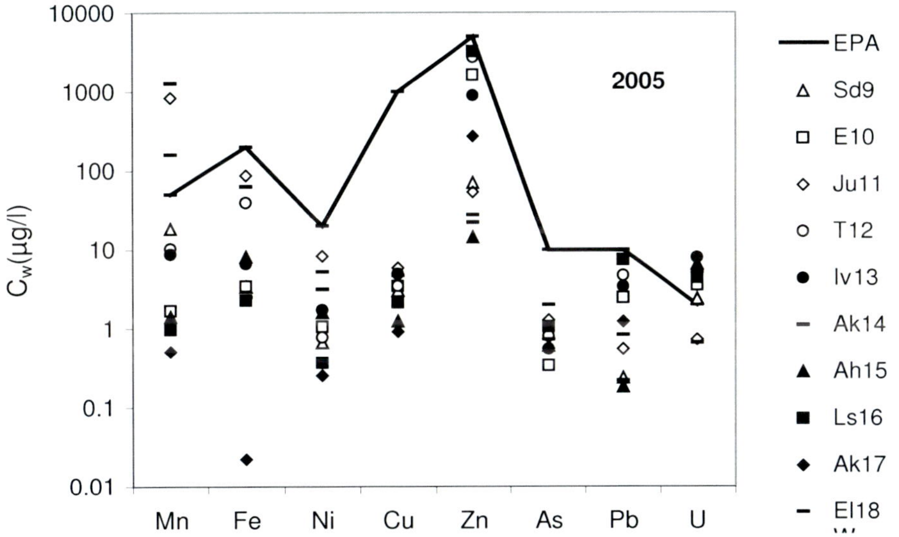

Appendix 3. Concentrations of heavy metals (µg/kg) bound to SPM in surface water and groundwater of the Akaki Sub-basin (April 2003).

Surface water	Fe	Cu	Zn	Mn	As	Pb	K	Ca	Ni
Ku1	74212.60	39.85	626.44	1199.71	10.25	0.00	539.33	8573.38	11.68
Dk4	5064.52	434.49	3853.05	12019.91	307.18	266.63	7325.07	74205.05	388.78
Dk5	16000.00	18.52	248.84	837.45	21.32	19.76	357.53	5013.01	16.00
Gr7	6186.94	5.34	197.97	221.01	1.36	0.00	114.77	533.99	8.16
Ju11	8360.73	4.46	70.03	137.47	1.45	0.00	59.67	961.31	0.98
AKt14	1432.64	2.53	17.43	53.70	0.84	1.17	32.34	329.89	1.71
El18	21.84	5.74	6.84	0.91	1.22	0.00	6.23	107.94	0.89

Groundwater	Fe	Cu	Zn	Mn	As	Pb	K	Ca	Ni
De2	35333.33	356.11	7464.56	960.09	2343.73	709.36	538.12	5464.14	61949.39
Dk3	36750.00	2613.49	15402.77	1197.74	919.90	838.37	965.01	12863.04	147729.64
De6	54266.67	5302.95	13775.08	1012.05	8033.76	910.14	1299.12	12904.32	120216.50
Be8	16363.64	4211.74	5010.53	711.72	592.71	558.06	678.85	4474.66	81850.34
Sd9	1622.50	68.33	440.83	51.22	69.28	124.56	72.97	698.33	8515.00
E10	138175.00	5845.81	37574.32	4363.04	0.00	10610.79	6216.03	59444.71	725200.08
Tl12	2515.00	18.23	518.13	65.00	164.72	40.63	39.73	378.13	4306.25
Iv13	1557.14	111.43	645.71	44.29	41.78	30.18	42.86	557.14	6168.57
Ah15	151373.00	187.00	2274.00	218.00	7809.63	194.97	196.94	3485.00	48091.00
Ls16	1787.00	172.00	447.00	35.50	257.03	51.00	46.00	404.50	3950.50
Ak17	394017.22	346.67	12594.44	476.67	13991.38	67.78	0.00	7415.56	34270.56
Td20	1391.43	67.14	379.29	60.71	70.65	115.00	67.86	593.57	7255.71
Am21	19800.00	57.14	1322.86	54.29	0.00	48.57	69.29	354.29	6755.00
Ab22	2968.33	155.00	398.33	138.33	0.00	118.33	121.67	776.67	14978.33

Appendix 4. Mineralogical compositions of the filter media (%).

Sample code	Quartz	Ortho-clase	albite	Clino Pyroxene	Calcite	Smectite	Kaolinite	Illite	Montmo-rillonite
KH1	2.00		-	0.00	0.00		93.00	7.00	
Wc2	8.00	0.80	3.00	-	3.00		-		8.00
Wl3	34.54		19.00	39.00	0.00		0.00	0.00	10.08
KVA	2.00	4.00	8.00	3.00	0.00		0.00	0.00	
DQS	2.00		13.00	11.00	0.00		0.00	0.00	
Yc6	22.00		41.00	14.00	0.00		0.00	0.00	
MnCw7	4.00	3.00	7.00	85.00	0.00		7.00	7.00	10.00
Ew(WB)Z8	0.00		5.00	-	0.00		0.00	0.00	23.00
BWK9	3.00	7.00		0.00	0.00		84.00	7.00	
DBWC10	14.00		27.00	9.00	0.00		17.00	0.00	
EBF	26.00		9.00	-	0.00		4.00	-	1.49
SVA	7.00	35.00		6.00	0.00	12.00	0.00	0.00	
Ew13	2.00	9.00	37.00	7.00	0.00		0.00	0.00	10.00

Mineralogical compositions of the filter media (%). *Contd.*

Sample code	Zeolite(Chabazite)	Clay fraction	Total	Hematite	CO2	S	Rest	Glass
KH1			22.80	0.00	0.11	0.05	-	_
Wc2		77.20	22.80	0.00	0.14	0.01	-	-
WI3		20.00	102.62	-	0.05	0.00	-	_
KVA		0.00	17.00	0.00	0.06	0.00	_	volcanic glass
DQS		10.00	26.00	10--20	0.09	0.01	-	_
Yc6		8.00	77.00	-	0.24	0.00	-	_
MnCw7		0.00	123.00	-	0.07	0.00	Fe-Oxide	-
Ew(WB)Z8	15.00	55.00	28.00		0.11	0.00	-	_
BWK9			101.00	-	0.13	0.00	Fe-Oxide 8%	_
DBWC10		-	67.00	-	0.18	0.00	33% Fe oxide+ organic matter	-
EBF		1.00	40.49	-	0.73	0.00	clay fraction	clay fraction
SVA		40.00	60.00	0.00	0.14	0.00	Fe-Oxide	-
Ew13		32.00	65.00	0.00	0.10	0.00	Fe-Oxide	_

*Clay minerals concentration could be higher than the measured ones. The real concentration measurements were hampered by poor crystallinity and interference effects of organic materials and possible amorphous phases like glass.

Appendix 5. Major elemental composition of the filter materials.

Sample	Na_2O	MgO	Al_2O_3	SiO_2	P_2O_5	CaO	K_2O	TiO_2	MnO	Fe_2O_3
	%	%	%	%	%	%	%	%	%	%
SVA	4.66	1.84	13.32	68.32	0.08	1.26	3.86	0.70	0.20	5.26
EWWB	2.44	2.76	19.22	62.88	0.04	3.00	2.36	0.51	0.13	5.83
YC	4.29	0.04	17.17	67.18	0.03		3.81	0.66	0.13	5.92
EW	1.04	4.26	18.60	51.93	0.23	4.19	0.83	2.08	0.18	15.09
DBWC	1.31	0.18	24.75	60.34	0.01	0.03	1.67	1.00	0.05	8.60
SVA	4.66	1.84	13.32	68.32	0.08	1.26	3.86	0.70	0.20	5.26
DQ scoria	1.95	9.57	15.69	45.54	0.38	8.58	0.58	1.99	0.18	12.58
EBF laterite	0.45	1.01	21.13	57.81	0.17	0.41	1.91	2.17	0.39	13.10
WC	0.11	0.02	39.60	47.41	0.02	0.07	0.74	0.10	0.05	1.36
MNCW	3.57	0.05	10.73	75.33	0.01	0.22	4.69	0.46	0.07	4.97

Appendix 6. Trace element contents of filter materials (ppm).

Sample	Ni	Cu	Zn	As	Br	Rb	Sr	Y	Zr	Nb	Mo
KVA vol. ash	5.21	226.21	233.99	7.93	4.08	119.31	8.47	94.74	838.05	123.49	3.79
KH	536.19	11.23	10.58	6.69	0.03	1.74	27.75	102.02	1549.05	212.62	4.89
BWK	395.90	20.86	34.60	6.68	0.00	63.22	8.82	70.92	55.01	60.63	0.50
WI	91.57	8.28	171.85	4.13	1.34	134.29	13.57	179.28	904.73	89.98	6.38
EWWBZ	41.44	68.96	108.99	0.76	1.86	24.08	3581.05	23.28	162.90	13.97	0.15
YC	82.47	16.03	126.16	12.01	0.00	72.45	14.98	131.09	966.75	128.23	6.11
EWWB	4.55	84.29	68.48	1.00	1.22	23.67	1398.45	8.63	377.20	42.38	1.78
DBWC	0.00	11.15	166.34	12.88	0.37	41.25	10.51	256.47	1539.39	214.64	4.95
SVA	1.03	10.95	194.85	7.17	3.20	96.31	104.73	80.76	716.71	93.44	4.31
DQ scoria	624.71	56.97	94.64	<5	<3	8.40	483.72	22.53	116.27	20.31	<5
WC	547.93	55.65	141.92	12.02	4.19	91.28	102.43	53.77	465.24	80.88	6.03
MNCW	704.45	97.54	98.04	<5	<3	39.38	125.13	45.81	268.11	63.70	<5
EBF laterite	480.96	53.90	160.13	8.83	4.53	107.59	48.22	51.12	693.22	102.78	6.83

Trace element contents of filter materials (ppm). *Contd.*

Sample	Ag	Cd	Sn	Sb	Ba	La	Ce	Pb
KVA vol. ash	0.32	0.22	6.85	0.98	71.47	110.40	187.08	11.81
KH	0.78	0.20	14.03	3.76	45.62	80.61	92.04	20.28
BWK	0.16	0.00	1.35	0.59	36.43	25.65	26.22	18.09
WI	0.56	0.54	7.57	1.51	173.11	116.23	187.20	0.00
EWWBZ	0.51	0.00	1.00	0.00	301.16	25.07	36.06	0.53
YC	0.38	0.32	8.37	0.00	450.70	109.24	193.02	1.75
EWWB	0.56	0.72	2.41	0.25	893.26	11.76	73.59	9.05
DBWC	0.59	0.00	10.83	0.00	158.35	161.09	300.47	0.00
SVA	0.85	0.24	5.08	0.02	262.38	83.20	177.48	5.10
DQ scoria	<1	<1	<2	<2	314.19	18.78	51.03	<5
WC	<1	<1	4.39	<2	485.25	83.96	224.01	15.35
MNCW	<1	<1	<2	<2	880.59	48.62	96.01	<5
EBF laterite	<1	<1	5.11	2.09	373.62	59.30	218.77	28.47

Appendix 7. Some selected heavy metal contents of the filter materials after sequential extraction in mg/kg (ZEIEN & BRÜMER 1989).

Sample	Fraction	Cr	Co	Ni	Cu	Zn
WI	Mobile	1.447	2.893	2.893	2.411	9.644
	Supplyable	0.964	0.000	0.000	1.929	4.340
	Mn oxide	1.447	5.304	0.000	1.929	21.217
	Organic	0.964	8.198	0.000	2.411	55.936
	Fe oxide (non cryst.)	1.929	10.126	0.482	1.447	15.913
	Fe oxide (cryst.)	2.893	13.502	6.751	3.375	183.238
	Total	9.644	40.023	10.126	13.502	290.288
	Fraction	Cr	Co	Ni	Cu	Zn
DQ scoria	Mobile	0.964	3.375	0.964	1.447	2.411
	Supplyable	0.964	0.000	0.000	1.929	0.964
	Mn oxide	1.447	8.680	0.000	2.893	2.411
	Organic	1.447	8.198	1.929	4.822	2.893
	Fe oxide (non cryst.)	28.932	26.039	41.470	18.806	24.110
	Fe oxide (cryst.)	48.221	27.486	34.719	16.395	18.806
	Total	81.975	73.778	79.082	46.292	51.596
	Fraction	Cr	Co	Ni	Cu	Zn
KVA vol. ash	Mobile	1.447	2.893	1.447	1.929	11.091
	Supplyable	0.964	4.822	0.000	1.929	1.447
	Mn oxide	1.447	0.482	0.000	1.929	3.858
	Organic	0.482	5.786	0.000	2.411	1.447
	Fe oxide (non cryst.)	2.411	14.466	0.000	1.929	3.858
	Fe oxide (cryst.)	2.411	13.984	0.000	1.929	18.806
	Total	9.162	42.434	1.447	12.055	40.505
	Fraction	Cr	Co	Ni	Cu	Zn
SVA	Mobile	1.447	3.375	3.858	2.893	11.091
	Supplyable	0.964	5.786	0.000	1.929	1.447
	Mn oxide	1.447	2.411	0.000	1.929	4.340
	Organic	0.964	10.126	0.000	1.447	6.751
	Fe oxide (non cryst.)	2.411	14.948	1.447	2.411	9.644
	Fe oxide (cryst.)	3.375	16.877	3.375	3.375	26.039
	Total	10.609	53.525	8.680	13.984	59.311

Some selected heavy metal contents of the filter materials after sequential extraction in mg/kg (ZEIEN & BRÜMER 1989). *Contd.*

Sample	Fraction	Cr	Co	Ni	Cu
MWB(d)	Mobile	1.447	6.269	2.411	0.964
	Supplyable	0.964	5.786	0.000	2.893
	Mn oxide	0.482	58.829	0.000	3.858
	Organic	0.964	24.110	7.233	8.198
	Fe oxide (non cryst)	50.632	21.217	29.415	14.948
	Fe oxide (cryst)	115.730	30.379	54.489	13.502
	Total	170.219	146.591	93.548	44.363
	Fraction	Cr	Co	Ni	Cu
EBF laterite	Mobile	0.964	4.822	3.858	2.411
	Supplyable	0.964	1.929	0.000	1.929
	Mn oxide	0.482	2.893	0.000	1.447
	Organic	1.447	7.715	0.000	2.893
	Fe oxide (non cryst)	2.893	11.573	1.447	2.411
	Fe oxide (cryst)	4.340	17.842	7.233	3.375
	Total	11.091	46.774	12.537	14.466
	Fraction	Cr	Co	Ni	Cu
EWWB	Mobile	0.964	6.269	6.751	2.893
	Supplyable	0.964	6.751	0.000	1.447
	Mn oxide	0.964	14.466	0.964	1.447
	Organic	0.482	13.502	0.000	3.858
	Fe oxide (non cryst)	4.822	17.842	4.340	5.786
	Fe oxide (cryst)	38.577	32.790	20.253	57.865
	Total	46.774	91.619	32.308	73.295

Appendix 8. Sorption of Cu, Ni, Zn, As, and Pb (µg/l) to one g/l of Wyoming type Na-Montmorillonite in simulated wastewater background solution.

Sample	Element	Spike conc	SPM bounded	Dissolved HM	SPM bounded(%)
SAW1	Cu	50	5.17	44.83	10.34
SAW2		100	38.38	61.62	38.38
SAW3		1000	510.20	489.80	51.02
SAW1	Ni	50	28.83	21.17	57.66
SAW2		100	46.93	53.07	46.93
SAW3		1000	360.40	639.60	36.04
SAW1	Zn	500	293.40	206.60	58.68
SAW2		1000	663.80	336.20	66.38
SAW3		1000	262.60	737.40	26.26
SAW1	As	50	12.47	37.53	24.94
SAW2		100	6.62	93.38	6.62
SAW3		1000	80.30	919.70	8.03
SAW1	Pb	50	47.79	2.22	95.57
SAW2		100	94.81	5.19	94.81
SAW3		1000	917.64	82.36	91.76

Appendix 9. Dissolved effluent concentrations of Mn, Fe, Ni, Cu, Zn, As, Cd and Pb (µg/l) over the 12 weeks experimental period.

a) KVA volcanic ash

t (day)	Mn	Fe	Ni	Cu	Zn	As	Cd	Pb
0	11.55	36.11	639.60	489.80	737.40	919.70	1.75	76.28
1	545.10	45.81	43.91	15.45	34.30	6.17	5.94	0.27
2	1486.00	12.70	64.65	45.96	7.76	8.30	2.67	0.35
4	392.80	23.84	65.76	77.80	47.83	183.20	0.93	1.91
8	44.45	17.45	126.90	46.72	12.61	2121.00	0.33	1.17
11	46.17	22.35	95.70	109.00	8.85	1251.00	0.39	0.61
15	23.08	13.06	205.90	39.80	24.56	1529.00	0.44	2.07
17	44.74	11.46	272.00	27.29	16.77	1588.00	0.23	1.41
22	99.17	8.91	349.50	15.64	20.78	1561.00	0.14	1.01
31	420.00	32.29	201.80	11.52	17.68	1350.00	0.12	0.78
36	404.10	33.94	170.30	8.38	30.61	1345.00	0.10	0.27
46	600.60	118.10	115.50	4.57	3.44	1339.00	0.18	6.06
50	460.90	14.85	64.76	8.45	11.89	961.40	0.15	12.16
58	418.60	22.13	61.06	6.93	12.95	974.30	0.50	4.63
71	232.90	26.85	41.47	4.09	3.67	1005.00	0.09	1.28
85	422.50	38.21	62.60	3.53	4.68	1159.00	0.36	0.89
95	395.50	39.67	40.05	3.29	5.96	955.50	0.18	0.49
106	333.10	41.06	33.41	8.42	20.19	1085.00	0.18	1.03

b) PYC pyroclastics

t (day)	Mn	Fe	Ni	Cu	Zn	As	Cd	Pb
0	11.55	36.11	639.60	489.80	737.40	919.70	1.75	76.28
1	1023.00	45.88	64.32	60.77	42.40	10.79	7.03	0.34
2	710.90	18.90	58.39	97.25	8.56	30.32	0.83	0.63
3	239.10	18.84	59.83	97.82	8.14	191.10	4.70	0.53
4	137.50	23.41	69.19	101.70	27.40	845.20	5.01	1.34
8	17.52	28.11	131.70	45.72	9.16	1985.00	0.45	0.78
11	25.96	16.81	103.00	165.30	8.01	1162.00	0.87	0.50
15	93.33	23.57	103.50	98.91	27.87	1421.00	0.91	1.69
17	35.86	12.41	179.60	30.27	10.50	1528.00	0.11	0.57
22	41.63	17.26	235.70	32.55	17.35	1437.00	0.14	0.83
31	211.20	12.73	246.20	53.53	14.94	1521.00	0.22	0.19
36	66.90	17.92	188.60	28.25	2.85	1206.00	0.05	0.29
44	96.96	3.96	144.00	80.68	2.48	1013.00	0.45	186.70
46	80.60	9.67	192.40	55.18	23.19	1329.00	0.23	196.80
50	38.08	13.10	184.20	9.24	2.38	884.10	0.17	4.30
58	109.00	27.89	151.50	3.58	10.10	710.80	0.38	4.06
71	62.32	39.91	101.50	2.60	2.67	885.50	0.07	3.95
85	55.88	25.70	104.30	7.71	2.76	1312.00	0.20	0.64
95	112.90	33.19	138.10	7.18	7.29	961.70	0.09	0.59
106	105.10	60.56	99.17	2.33	6.32	1707.00	0.07	0.57

c) MnR sediment

t (day)	Mn	Fe	Ni	Cu	Zn	As	Cd	Pb
0	11.55	36.11	639.60	489.80	737.40	919.70	1.75	76.28
1	9.83	1.86	0.92	1.59	5.10	3.81	0.17	0.24
2	12.36	0.99	3.56	2.75	1.83	2.16	0.06	0.39
3	24.21	1.42	5.03	2.91	5.12	3.59	1.18	0.46
8	1.66	1.07	13.84	7.82	13.28	10.69	0.14	0.68
11	1.38	1.78	7.59	5.51	5.35	5.97	0.14	0.17
15	2.93	2.10	4.87	4.85	9.27	10.67	0.14	0.52
17	2.59	3.07	6.55	5.63	6.60	10.86	0.06	0.31
22	4.52	1.53	9.69	6.29	7.71	16.87	0.10	0.35
31	11.93	1.08	8.24	4.54	2.33	18.08	0.12	0.12
36	6.91	1.12	7.84	4.61	1.53	17.82	0.03	0.12
44	15.45	4.18	4.06	4.40	3.54	18.85	0.17	144.50
46	9.33	2.65	10.81	6.65	10.54	36.55	0.07	139.80
50	18.73	1.57	7.42	4.50	1.61	19.16	0.23	6.41
58	11.50	2.43	8.25	7.33	8.50	15.90	0.12	1.53
71	203.50	3.69	6.82	2.90	3.84	18.54	0.12	2.78
85	21.74	1.91	6.85	4.27	5.76	17.39	0.15	0.68
95	55.31	2.34	5.68	2.79	3.94	15.10	0.09	0.43
106	53.14	1.41	6.72	4.01	2.21	15.76	0.08	0.19

d) DQ scoria

t (day)	Mn	Fe	Ni	Cu	Zn	As	Cd	Pb
0	11.55	36.11	639.60	489.80	737.40	919.70	1.75	76.28
1	124.30	6.47	13.33	16.37	7.59	6.36	1.04	0.52
2	157.00	8.59	26.89	31.30	2.96	8.19	0.78	0.44
3	169.80	10.05	29.80	36.29	2.89	10.65	2.20	0.45
4	118.30	12.11	34.45	41.83	6.03	11.78	0.96	0.75
8	516.10	13.98	59.48	31.40	7.47	684.30	0.33	0.74
11	571.80	22.52	50.39	74.79	8.34	596.30	0.45	0.57
15	721.90	14.76	67.15	37.02	15.46	836.20	0.41	1.24
22	881.70	33.02	93.73	27.86	16.35	773.00	0.17	0.90
31	1592.00	78.77	89.48	11.76	4.97	590.00	0.12	0.74
44	822.00	33.92	78.14	7.78	5.86	1119.00	0.21	193.40
46	1193.00	36.61	72.75	7.83	14.11	878.60	0.12	1137.00
50	763.70	18.33	80.33	11.80	19.90	925.40	0.77	4.43
58	618.70	16.09	65.91	9.06	9.06	815.50	0.11	1.68
71	550.30	36.65	45.92	1.80	2.73	684.70	0.04	0.90
85	474.80	7.30	54.89	4.01	3.11	950.80	0.11	0.32
95	1005.00	111.00	43.55	1.10	4.57	889.50	0.06	0.39
106	558.70	24.63	110.70	3.21	4.10	1591.00	0.03	0.24

e) DQ scoria (< 0.8 mm)

t (day)	Mn	Fe	Ni	Cu	Zn	As	Cd	Pb
0	11.55	36.11	639.60	489.80	737.40	919.70	1.75	76.28
1	248.00	16.82	4.13	5.50	3.85	4.58	1.32	0.27
2	210.00	4.86	14.67	15.05	2.81	4.51	0.46	0.43
4	74.67	8.99	26.86	27.58	17.02	5.33	0.79	0.84
8	424.30	13.73	43.99	27.86	12.06	23.10	0.42	0.69
11	1255.00	20.35	40.76	46.28	8.52	20.01	0.47	0.39
15	774.30	18.59	45.04	24.12	12.73	92.41	0.12	0.99
17	567.10	70.49	63.35	19.20	14.41	206.60	0.08	0.80
22	543.50	44.23	59.02	15.37	10.17	122.00	0.12	0.75
31	1596.00	82.73	57.67	11.54	7.74	525.30	0.07	0.53
36	1458.00	78.09	48.88	4.27	3.74	687.70	0.07	0.25
46	1546.00	72.80	47.97	6.45	10.67	539.60	0.36	2883
50	1228.00	31.26	38.94	6.75	11.10	661.70	0.71	26.24
58	1775.00	37.70	41.68	4.59	9.10	723.70	0.14	6.21
71	1280.00	17.03	31.38	3.46	6.77	630.10	0.05	2.30
85	2135.00	65.82	31.28	3.72	5.11	1166.00	0.18	1.10
95	1321.00	31.52	30.66	3.27	4.91	1112.00	0.07	0.67
106	1175.00	47.29	26.46	1.56	2.51	1340.00	0.05	0.27

f) EBF laterite

t (day)	Mn	Fe	Ni	Cu	Zn	As	Cd	Pb
0	11.55	36.11	639.60	489.80	737.40	919.70	1.75	76.28
1	1234.00	10.98	44.30	2.89	38.41	1.53	2.18	0.58
2	2252.00	9.13	75.94	5.80	31.44	1.39	1.28	0.35
3	1898.00	4.33	44.96	17.18	4.71	1.70	1.57	0.40
4	1039.00	4.50	12.02	21.48	4.71	1.93	2.31	0.50
8	211.20	5.33	38.66	28.99	9.81	8.40	0.20	0.73
11	1395.00	6.51	48.41	55.93	8.40	9.04	0.23	0.37
15	902.50	9.04	51.64	20.91	15.67	18.03	0.21	0.97
17	1393.00	45.60	62.71	18.28	8.88	23.91	0.12	0.78
22	1880.00	60.08	70.07	15.05	13.31	58.39	0.13	1.01
31	1143.00	54.69	66.95	6.30	4.40	322.40	0.08	0.79
36	917.30	33.69	59.20	3.86	2.55	597.90	0.04	0.34
46	458.80	18.93	53.93	4.91	4.26	423.00	0.12	66.61
50	1512.00	42.86	42.72	4.60	3.08	568.60	0.24	2.08
58	1310.00	83.13	36.71	6.32	3.04	793.10	0.08	1.60
71	1181.00	56.84	45.40	4.51	7.28	655.20	0.05	0.86
85	1426.00	10.46	35.12	2.60	7.49	1751.00	0.16	0.52
95	1444.00	22.50	33.21	5.41	7.18	918.90	0.07	0.30
106	1112.00	29.16	52.03	2.63	4.75	1135.00	0.04	0.35

g) QTZ sand

t (day)	Mn	Fe	Ni	Cu	Zn	As	Cd	Pb
0	11.55	36.11	639.60	489.80	737.40	919.70	1.75	76.28
1	426.10	544.90	118.60	47.13	50.86	19.42	5.82	0.32
2	157.60	19.67	124.30	64.25	3.68	106.00	0.42	0.72
3	88.75	19.09	110.60	69.45	3.28	333.40	2.27	0.72
4	47.20	24.27	72.36	79.77	56.09	707.90	2.00	1.05
8	16.44	17.94	158.00	30.79	12.67	1661.00	0.26	0.88
11	10.47	25.91	134.60	35.49	12.18	1144.00	0.20	0.75
15	11.91	17.57	188.50	30.26	10.09	1500.00	0.18	1.18
17	16.14	19.04	211.20	26.43	9.98	1430.00	0.09	1.16
22	20.33	16.60	226.10	32.35	15.68	1303.00	0.14	1.01
31	36.59	20.04	215.50	18.81	5.44	1162.00	0.09	1.09
36	43.83	14.36	182.70	8.69	3.63	919.70	0.04	0.85
44	34.08	26.25	114.00	7.58	9.38	867.30	0.05	14.98
46	34.53	38.00	115.80	6.42	4.55	914.90	0.72	25.46
50	31.04	20.45	112.50	14.62	4.23	767.30	0.29	2.17
58	46.31	13.92	100.00	8.42	4.18	882.00	0.07	1.05
71	33.35	32.27	111.00	7.35	3.10	1117.00	0.04	0.66
85	59.31	9.86	473.80	17.48	5.34	1592.00	0.19	0.51
95	44.70	21.98	408.00	12.11	20.20	3741.00	0.12	0.97
106	50.16	42.32	128.40	6.88	6.73	1555.00	0.05	0.79

Appendix 10. Desorption of heavy metals from the used media after a modified Leaching test (DIN 38414-S4) (mg/kg).

Sample	Al	Cr	Mn	Fe	Ni	Cu	Zn	As	Mo	Cd	Pb
KVA	58.79	0.05	18.50	40.15	0.74	0.79	2.35	0.11	0.21	0.14	0.08
KVA0-5	63.34	0.13	6.42	37.02	58.14	31.12	37.42	65.08	1.14	0.22	2.58
KVA5-15	37.26	0.07	11.88	35.55	18.59	5.38	11.31	12.44	0.28	0.09	1.06
KVA 15-25	43.38	0.12	19.51	78.72	10.83	1.62	3.63	4.90	0.26	0.05	0.31
KVA25-30	57.54	0.10	16.15	59.81	4.59	1.09	3.96	4.64	0.12	0.06	0.28
PYC0-10	15.15	0.03	7.58	6.07	24.52	7.70	15.69	26.34	0.24	0.06	0.34
PYC0-20	17.59	0.04	9.87	6.41	55.96	18.85	27.37	40.90	0.71	0.11	0.56
PYC20-30	178.13	0.13	8.65	75.96	6.87	5.11	7.67	4.46	0.09	0.06	0.46
DQ	41.18	1.09	2.97	18.08	0.31	0.40	1.17	0.06	0.12	0.24	0.08
DQ0-5	39.56	0.04	37.68	16.24	44.14	27.40	12.35	72.12	0.58	0.08	0.61
DQ5-15	13.82	0.04	58.94	8.61	45.14	55.62	10.75	54.70	0.40	0.06	0.35
DQ15-25	7.81	0.04	19.61	5.19	16.21	9.90	4.77	21.82	0.30	0.06	0.16
DQ25-30	20.84	0.06	26.68	12.87	3.36	1.49	2.41	10.66	0.12	0.05	0.12
EBF	157.00	1.83	66.94	122.74	1.20	0.30	0.87	0.06	0.10	0.37	0.08
EBF0-15	21.54	0.05	175.98	18.16	33.40	20.78	5.35	44.88	0.46	0.06	0.25
EBF15-25	22.30	0.05	156.58	23.38	9.28	2.93	2.10	19.52	0.20	0.09	0.19
EBF25-30	14.46	0.04	145.80	11.59	1.49	1.05	0.87	13.42	0.15	0.04	0.06
QTZ0-5	5.92	0.03	4.75	3.67	46.18	16.89	21.90	33.38	0.54	0.07	0.54
QTZ5-15	8.45	0.03	7.05	4.44	36.42	17.33	15.69	26.26	0.32	0.06	0.48
QTZ15-30	4.99	0.02	7.48	2.56	28.42	7.30	14.17	26.52	0.26	0.05	0.26
MnR0-5	8.47	0.03	564.20	4.50	7.71	13.08	1.55	23.60	7.23	0.04	0.19
MnR5-15	6.62	0.09	48.08	3.57	1.10	2.84	1.20	15.33	9.32	0.05	0.11
MnR15-25	9.65	0.12	68.74	5.00	0.25	0.42	1.12	1.25	0.94	0.05	0.09
MnR25-30	1.81	0.13	6.11	1.59	0.21	0.44	1.29	0.22	0.25	0.04	0.08

Appendix 11. Comparison of the BCR three-stage sequential extraction results for evaluating fractionation of heavy metals in the column sludge (mg/kg).

Sample	Fraction	Ni	Cu	Zn	As	Pb	Mo	Cd	Al	Cr	Fe	Mn
KVA 0-5	Exchangeable	1027.22	1737.76	2866.33	1044.80	578.52	4.19	6.07	378.42	114.55	2948.89	85.26
KVA 5-15	Exchangeable	282.64	261.54	317.40	229.46	74.16	3.70	10.97	521.30	112.26	2030.06	101.73
KVA 15-25	Exchangeable	161.18	64.93	37.95	99.98	16.22	4.13	0.88	1454.58	145.20	1939.15	95.56
KVA 25-30	Exchangeable	80.21	29.55	20.78	81.24	8.15	3.53	0.71	1084.26	134.26	2096.61	77.66
KVA 0-5	Reducible	339.02	439.98	980.92	61.27	479.58	20.91	2.03	1927.66	4.82	2590.37	43.55
KVA 5-15	Reducible	46.78	42.60	58.28	31.43	59.60	9.00	0.33	881.65	4.58	1339.39	23.63
KVA 15-25	Reducible	21.63	26.54	30.38	25.66	13.49	8.36	0.30	383.19	4.82	950.85	9.54
KVA 25-30	Reducible	15.27	20.08	22.43	22.43	8.24	5.84	0.19	547.78	4.62	1054.48	6.41
KVA 0-5	Oxidizable	44.08	67.39	88.34	14.90	72.78	12.90	0.14	3270.33	38.56	1823.03	31.60
KVA 5-15	Oxidizable	16.28	13.18	26.33	9.26	24.28	9.17	0.02	1092.00	47.41	790.55	19.49
KVA 15-25	Oxidizable	8.16	9.34	17.40	8.72	8.16	7.61	0.02	699.17	41.22	890.14	13.22
KVA 25-30	Oxidizable	4.85	9.75	16.40	6.52	5.98	7.28	0.02	625.78	44.37	700.19	19.60
Pycd 0-10	Exchangeable	652.97	938.87	1302.02	711.42	294.16	3.97	1.65	846.54	119.84	3137.44	85.68
Pycd 10-20	Exchangeable	295.93	359.07	434.42	317.87	102.52	3.20	0.75	1194.17	136.98	2749.49	66.60
Pycd 20-30	Exchangeable	99.88	60.92	10.50	49.77	6.50	2.92	0.77	2147.27	122.08	1686.38	51.96
Pycd 0-10	Reducible	161.81	168.33	338.10	27.95	240.46	4.64	0.48	2867.67	3.93	2113.77	47.42
Pycd 10-20	Reducible	37.83	45.20	50.97	21.85	56.73	4.17	0.29	2624.55	3.82	1909.51	55.50
Pycd 20-30	Reducible	14.15	19.62	17.31	13.31	6.76	2.69	0.25	2140.89	2.69	999.08	34.62
Pycd 0-10	Oxidizable	18.57	24.25	33.71	5.06	14.42	5.79	0.02	3303.05	41.33	607.30	15.14
Pycd 10-20	Oxidizable	6.70	14.79	15.19	4.72	7.06	7.52	0.02	2434.38	42.38	505.59	13.50
Pycd 20-30	Oxidizable	5.23	24.85	14.88	5.82	4.23	5.75	0.02	1978.12	37.44	546.77	35.50
Pyc 0-10	Exchangeable	782.28	983.08	1295.11	771.68	338.11	3.95	1.77	928.37	150.79	3511.76	58.61
Pyc 10-20	Exchangeable	579.89	649.23	912.30	698.63	230.24	3.10	1.03	2559.23	143.12	3943.37	158.57
Pyc 20-30	Exchangeable	218.20	162.68	183.14	193.09	56.65	2.82	0.75	2486.72	146.53	2710.03	73.08
Pyc 0-10	Reducible	164.82	161.38	296.92	24.45	165.73	2.79	0.51	2815.17	3.63	1912.99	34.21
Pyc 10-20	Reducible	62.15	78.32	106.34	23.35	92.18	2.51	0.31	5634.19	3.59	1839.77	58.02
Pyc 20-30	Reducible	18.40	27.32	25.63	14.08	23.51	2.16	0.23	3149.39	3.19	1354.08	38.67
Pyc 0-10	Oxidizable	24.22	31.79	48.67	4.94	13.71	6.04	0.02	3062.80	48.46	758.95	22.13
Pyc 10-20	Oxidizable	15.58	33.37	27.89	5.63	10.60	6.21	0.02	6156.82	45.65	897.07	23.89
Pyc 20-30	Oxidizable	7.09	18.02	16.38	4.62	6.99	6.15	0.02	3137.33	52.12	656.53	19.29

Comparison of the BCR three-stage sequential extraction results for evaluating fractionation of heavy metals in the column sludge (mg/kg). *Contd.*

Sample	Fraction	Ni	Cu	Zn	As	Pb	Mo	Cd	Al	Cr	Fe	Mn
DQ 0-5	Exchangeable	1774.55	1369.75	2535.28	866.05	309.00	3.86	4.20	6588.42	149.62	10851.55	965.65
DQ 5-15	Exchangeable	977.93	1057.86	1573.27	774.92	265.20	3.50	1.27	3321.35	150.32	9521.72	830.85
DQ15-25	Exchangeable	568.76	432.69	625.60	407.75	115.14	3.21	0.93	3562.43	151.98	12555.36	792.38
DQ 25-30	Exchangeable	188.88	83.31	89.80	109.74	18.28	3.17	0.85	3467.23	154.16	5357.58	629.77
DQ 0-5	Reducible	291.24	233.46	395.57	184.27	216.43	10.05	1.08	10373.41	21.37	23649.23	443.78
DQ 5-15	Reducible	152.42	139.02	213.21	91.98	146.98	4.82	0.40	6575.59	12.96	13433.78	289.51
DQ 15-25	Reducible	109.69	64.85	88.11	83.18	84.67	3.54	0.32	4864.66	15.72	16395.61	243.77
DQ25-30	Reducible	54.85	25.20	34.29	32.30	14.34	2.46	0.32	2488.57	10.99	9697.83	178.84
DQ 0-5	Oxidizable	89.64	51.29	61.44	13.48	83.69	4.04	0.02	9073.30	64.49	4398.30	157.57
DQ 5-15	Oxidizable	39.94	32.41	40.43	6.76	38.32	4.93	0.02	6093.64	46.76	2872.59	105.12
DQ 15-25	Oxidizable	30.33	18.70	26.61	6.90	22.14	3.90	0.02	3742.09	54.27	3422.22	65.73
DQ 25-30	Oxidizable	16.91	11.27	16.53	3.24	7.72	3.07	0.02	1731.79	58.94	2200.77	50.63
EBF 0-15	Exchangeable	1111.66	1154.30	1427.13	856.17	353.54	3.01	1.95	1739.76	143.41	9520.14	2143.79
EBF 15-25	Exchangeable	505.43	402.67	489.23	277.88	110.60	2.92	1.10	1529.48	157.05	5448.99	1727.41
EBF 25-30	Exchangeable	145.16	56.93	35.08	125.18	20.85	2.65	0.82	2211.72	164.87	10798.36	1686.54
EBF 0-15	Reducible	216.53	192.96	336.41	193.98	229.74	2.29	0.56	4623.64	10.00	15159.81	656.38
EBF 15-25	Reducible	77.66	81.13	101.84	123.11	110.92	1.73	0.40	3986.87	10.31	13830.99	486.49
EBF 25-30	Reducible	17.61	27.19	37.23	78.65	21.05	1.46	0.21	2493.66	7.66	25826.64	1148.18
EBF 0-15	Oxidizable	58.33	48.28	66.98	14.51	66.26	5.05	0.02	6590.93	50.21	2725.94	94.23
EBF 15-25	Oxidizable	17.52	17.93	30.98	9.14	30.11	5.03	0.02	4650.94	57.61	2002.78	67.21
EBF 25-30	Oxidizable	3.79	9.35	17.47	5.54	10.17	5.47	0.02	2606.66	44.46	1600.23	45.03
Qtz 0-5	Exchangeable	782.99	1104.29	1278.73	664.74	349.37	3.58	2.11	511.12	163.32	3338.84	43.99
Qtz 5-15	Exchangeable	598.79	771.23	831.54	543.48	233.92	3.04	1.17	437.25	155.60	3155.55	46.29
Qtz 15-30	Exchangeable	545.58	544.56	675.13	553.56	192.69	2.47	0.93	403.86	143.66	3028.46	48.13
Qtz 0-5	Reducible	196.83	206.72	438.03	37.52	198.97	9.60	0.74	1228.17	3.72	1875.97	24.59
Qtz 5-15	Reducible	121.00	117.64	213.95	24.50	126.80	4.30	0.30	1037.48	3.83	1804.75	9.73
Qtz 15-30	Reducible	107.86	90.67	172.65	31.30	104.17	2.51	0.26	1162.63	4.05	2046.47	11.68
Qtz 0-5	Oxidizable	20.13	32.04	60.64	3.74	16.41	4.58	0.02	1392.48	45.29	584.94	9.01
Qtz 5-15	Oxidizable	7.81	11.78	19.59	2.54	6.36	3.23	0.02	762.48	43.37	377.45	5.28
Qtz 15-30	Oxidizable	6.57	11.72	17.68	2.73	6.73	2.86	0.02	791.42	38.71	429.93	5.11

Comparison of the BCR three-stage sequential extraction results for evaluating fractionation of heavy metals in the column sludge (mg/kg). *Contd.*

Sample	Fraction	Ni	Cu	Zn	As	Pb	Mo	Cd	Al	Cr	Fe	Mn
Dq08 0-5	Exchangeable	2067.14	2498.32	3818.32	1167.10	425.15	2.85	4.90	6337.80	144.59	14309.72	1367.39
Dq08 5-15	Exchangeable	328.41	172.03	205.22	243.69	15.52	3.00	1.16	9546.43	173.51	18499.28	1569.74
Dq08 15-30	Exchangeable	1011.82	849.54	1066.44	514.59	102.64	2.46	1.40	7757.28	146.33	20380.74	1687.25
Dq08 0-5	Reducible	394.72	316.26	532.52	251.05	427.56	2.10	1.08	11386.13	28.50	36510.48	645.01
Dq08 5-15	Reducible	135.00	48.58	76.67	87.05	20.36	2.15	0.77	7544.80	24.78	34932.01	579.89
Dq08 15-30	Reducible	230.55	135.29	176.95	157.60	133.65	1.01	0.33	9908.58	29.49	36783.18	627.29
Dq08 0-5	Oxidizable	95.81	57.92	57.75	18.20	120.83	2.27	0.02	9721.71	52.62	6330.68	120.70
Dq08 5-15	Oxidizable	38.56	15.57	23.66	5.83	14.49	2.70	0.02	3832.98	52.40	4335.12	100.91
Dq08 15-30	Oxidizable	56.33	30.58	30.16	9.26	44.94	2.18	0.02	6347.43	48.79	3426.39	90.74
MnR 0-5	Exchangeable	2443.44	2102.83	3833.68	2348.73	15.63	0.36	18.04	11126.62	115.20	1228.99	2443.44
MnR 5-15	Exchangeable	565.50	321.67	1239.68	625.90	1.22	0.36	6.76	7731.98	124.73	240.90	565.50
MnR 15-20	Exchangeable	15.73	8.61	11.16	85.44	0.09	0.36	0.18	12114.94	120.81	16.59	15.73
MnR 20-30	Exchangeable	10.62	7.33	7.38	23.79	0.19	0.38	0.19	13150.15	129.23	21.15	10.62
MnR 0-5	Reducible	555.97	778.85	913.08	1854.53	167.32	7.25	14.41	12499.77	32.36	30766.79	555.97
MnR 5-15	Reducible	635.05	1027.21	1519.82	801.35	44.68	31.62	33.24	17523.42	16.13	43700.90	635.05
MnR 15-20	Reducible	508.43	783.93	2256.05	132.78	6.81	33.17	51.46	19826.97	12.01	52818.72	508.43
MnR 20-30	Reducible	382.19	618.47	1942.65	41.09	3.57	23.60	47.67	22912.75	9.12	61470.48	382.19
MnR 0-5	Oxidizable	9.79	37.12	174.16	18.60	135.41	5.29	0.91	892.23	34.65	709.87	1313.83
MnR 5-15	Oxidizable	40.54	79.37	54.41	34.86	499.64	16.76	0.81	4828.47	68.83	1428.38	1665.77
MnR 15-20	Oxidizable	1.93	34.52	24.83	14.31	85.31	34.45	0.02	6872.87	36.87	1098.22	3494.49
MnR 20-30	Oxidizable	0.99	28.63	22.05	6.55	48.23	33.36	0.02	6309.23	39.47	915.62	3594.54

In der Reihe "Karlsruher Geochemische Hefte" (ISSN 0943-8599) sind erschienen:

Band 1: U. Kramar (1993)
Methoden zur Interpretation von Daten der geochemischen Bachsedimentprospektion am Beispiel der Sierra de San Carlos/ Tamaulipas Mexiko

Band 2: Z. Berner (1993)
S-Isotopengeochemie in der KTB - Vorbohrung und Beziehungen zu den Spurenelementmustern der Pyrite

Band 3: J.-D. Eckhardt (1993)
Geochemische Untersuchungen an jungen Sedimenten von der Galapagos-Mikroplatte:
Hydrothermale und stratigraphisch signifikante Signale

Band 4: B. Bergfeldt (1994)
Lösungs- und Austauschprozesse in der ungesättigten Bodenwasserzone und Auswirkungen auf das Grundwasser

Band 5: M. Hodel (1994)
Untersuchungen zur Festlegung und Mobilisierung der Elemente As, Cd, Ni und Pb an ausgewählten Festphasen unter besonderer Berücksichtigung des Einflusses von Huminstoffen.

Band 6: T. Bergfeldt (1995)
Untersuchungen der Arsen- und Schwermetallmobilität in Bergbauhalden und kontaminierten Böden im Gebiet des Mittleren Schwarzwaldes.

Band 7: M. Manz (1995)
Umweltbelastungen durch Arsen und Schwermetalle in Böden, Halden, Pflanzen und Schlacken ehemaliger Bergbaugebiete des Mittleren und Südlichen Schwarzwaldes.

Band 8: J. Ritter (1995)
Genese der Mineralisation Herrmanngang im Albtalgranit (SE-Schwarzwald) und Wechselwirkungen mit dem Nebengestein.

Band 9: J. Castro (1995)
Umweltauswirkungen des Bergbaus im semiariden Gebiet von Santa Maria de la Paz, Mexiko.

Band 10: T. Rüde (1996)
Beiträge zur Geochemie des Arsens.

Band 11: J. Schäfer (1998)
Einträge und Kontaminationspfade Kfz-emittierter Platin-Gruppen-Elemente (PGE) in verschiedenen Umweltkompartimenten.

Band 12: M. A. Leosson (1999)
A Contribution to the Sulphur Isotope Geochemistry of the Upper Continental Crust: The KTB Main Hole - A Case Study

Band 13: B. A. Kappes (2000)
Mobilisierbarkeit von Schwermetallen und Arsen durch saure Grubenabwässer in Bergbaualtlasten der Ag-Pb-Zn-Lagerstätte in Wiesloch

Band 14: H. Philipp (2000)
The behaviour of platinum-group elements in petrogenetic processes:
A case study from the seaward-dipping reflector sequence (SDRS), Southeast Greenland volcanic rifted margin

Band 15: E. Walpersdorf (2000)
Nähr- und Spurenelementdynamik im Sediment/Wasser-Kontaktbereich nach einer Seekreideaufspülung - Pilotstudie Arendsee -

Band 16: R. H. Kölbl (2000)
Models of hydrothermal plumes by submarine diffuse venting in a coastal area: A case study for Milos, South Aegean Volcanic Arc, Greece

Band 17: U. Heiser (2000)
Calcium-rich Rhodochrosite layers in Sediments of the Gotland Deep, Baltic Sea, as Indicators for Seawater Inflow

In der Fortsetzungsreihe "Karlsruher Mineralogische und Geochemische Hefte" (ISSN 1618-2677) sind bisher erschienen:

Band 18: S. Norra (2001)
Umweltgeochemische Signale urbaner Systeme am Beispiel von Böden, Pflanzen, und Stäuben in Karlsruhe

Band 19: M. von Wagner, Alexander Salichow, Doris Stüben (2001)
Geochemische Reinigung kleiner Fließgewässer mit Mangankiesen, einem Abfallsprodukt aus Wasserwerken (GReiFMan)

Band 20: U. Berg (2002)
Die Kalzitapplikation als Restaurierungsmaßnahme für eutrophe Seen – ihre Optimierung und Bewertung

Band 21: Ch. Menzel (2002)
Bestimmung und Verteilung aquatischer PGE-Spezies in urbanen Systemen

Band 22: P. Graf (2002)
Meta-Kaolinit als latent-hydraulische Komponente in Kalkmörtel

Band 23: D. Buqezi-Ahmeti (2003)
Die Fluidgehalte der Mantel-Xenolithe des Alkaligesteins-Komplexes der Persani-Berge, Ostkarpaten, Rumänien: Untersuchungen an Fluid-Einschlüssen

Band 24: B. Scheibner (2003)
Das geochemische Verhalten der Platingruppenelemente bei der Entstehung und Differenzierung der Magmen der Kerguelen-Flutbasaltprovinz (Indischer Ozean)

Ab Band 25 erscheinen die Karlsruher Mineralogischen und Geochemischen Hefte im Karlsruher Universitätsverlag online unter der Internetadresse

http://www.uvka.de/univerlag/institut.php?fakultaet=1

Auf Wunsch sind beim Karlsruher Universitätsverlag gedruckte Exemplare erhältlich („print on demand").

Band 25: A. Stögbauer (2005)
Schwefelisotopenfraktionierung in abwasserbelasteten Sedimenten - Biogeochemische Umsetzungen und deren Auswirkung auf den Schwermetallhaushalt

Band 26: X. Xie (2005)
Assessment of an ultramicroelectrode array (UMEA) sensor for the determination of trace concentrations of heavy metals in water

Band 27: F. Friedrich (2005)
Spectroscopic investigations of delaminated and intercalated phyllosilicates

Band 28: L. Niemann (2005)
Die Reaktionskinetik des Gipsabbindens: Makroskopische Reaktionsraten und Mechanismen in molekularem Maßstab

Band 29: F. Wagner (2005)
Prozessverständnis einer Naturkatastrophe: eine geo- und hydrochemische Untersuchung der regionalen Arsen-Anreicherung im Grundwasser West-Bengalens (Indien)

Band 30: F. Pujol (2005)
Chemostratigraphy of some key European Frasnian-Famennian boundary sections (Germany, Poland, France)

Band 31: Y. Dikikh (2006)
Adsorption und Mobilisierung wasserlöslicher Kfz-emittierter Platingruppenelemente (Pt, Pd, Rh) an verschiedenen bodentypischen Mineralen

Band 32: F. I. Müller (2007)
Influence of cellulose ethers on the kinetics of early Portland cement hydration

Made in United States
North Haven, CT
01 July 2026

13772511R00146